Teaching the Metric System

With Activities

THE PROFESSIONAL EDUCATION SERIES

Teaching the Metric System

With Activities

by

FRANCIS E. MASAT
Associate Professor
of Mathematics
Glassboro State College

and

CHARLES H. PAGE
Associate Professor
of Elementary Education
Glassboro State College

PROFESSIONAL EDUCATORS PUBLICATIONS, INC.
LINCOLN, NEBRASKA

Library of Congress Catalog Card No.: 76–16273

ISBN 0–88224–100–1

Contents

Preface

The Workshop in Elementary School Mathematics, sponsored by the Department of Elementary Education and the Department of Mathematics at Glassboro State College, decided that since elementary teachers and mathematics coordinators were seeking help in the area of metrication, it would assemble the information and ideas from its workshops into a book on metrics. As most elementary mathematics texts do not currently provide adequate material and activities for metrication, the teachers participating in the 1974 and 1975 workshops and the authors put together basic information, activities, and procedures on making the metric system easier to teach and to learn. What has emerged is a simplified, concise, and self-contained metric book. While the work is primarily designed for in-service elementary teachers, it has been successfully used as a supplement in courses for prospective teachers.

Chapter 1 provides a brief history of the metric system, while Chapter 2 explains in a simplified and concise manner how the metric system works. Examples and recommended usages are also included. Chapter 3 gives practical advice and procedures for teaching the metric system. One of the most important responsibilities of the teacher will be to instill the fundamental principle of "thinking metric." The "think metric" concept is stressed throughout, as it is the idea of "thinking metric" that is most important.

Chapter 4 constitutes the major thrust of the book. In particular, students learn metric measurements by using them, i.e., confusing conversions from our old American system to the metric system are minimized. Concepts and skills are taught and experienced through a variety of activities, including measuring, solving realistic problems, puzzles, games, approximating, and physical activities. Many of the activities included have been field tested and refined in the elementary classroom. The activities are also designed so as to lend themselves to being applied at various levels. Additionally, the activities can be used to supplement elementary mathematics programs. Appendixes containing a large reference section on metric sources and sections of supplementary material complete the presentation.

The immediate response from most adults and students is to approach the metric system by trying to memorize lists of numerical conversions and equivalences between the two systems. However, the real secret to learning the metric system easily and quickly is to learn to "think metric." Real difficulties in learning the metric system will occur if we try to convert everything back to the American system. For the most part, we simply won't need to convert between the two systems.

Acknowledgments

The authors are indebted to Ms. Joanne Thompson and Ms. Anne Savin for patiently typing and retyping the many revisions this book has undergone.

A special thanks goes to Mr. Robert Simons of the Curriculum Development Council for Southern New Jersey for his many insights, suggestions, and encouragement. Acknowledgment is also made to the National Bureau of Standards, Washington, D.C., for checking and commenting on the entire text, and the following members of the workshop:

Don Becker
Glen Landing School
Blackwood, N.J.

James B. Chandler
St. John's Day School
Millville, N.J.

Susan L. Clark
Jaggard School
Marlton, N.J.

Carol DeRuyter
Arthur Rann School
Absecon, N.J.

Loris Grunow
Arthur Rann School
Absecon, N.J.

Dennis Jones
Bell Oaks School
Bellmawr, N.J.

Sr. Barbara Kardos
St. Mathews Catholic School
East Syracuse, N.Y.

Florence Moore
St. James Regional Elem. School
Penns Grove, N.J.

Karen Sanderson
East Greenwich Twp. Elem. School
Mickleton, N.J.

Marie Sangarlo
Longfellow School
Pennsauken, N.J.

Don Weatherby
Logan Township School
Bridgeport, N.J.

Finally, the authors are indebted to the series editors for their helpful aid and assistance.

CHAPTER 1

The Development of the Metric System

THE DEVELOPMENT OF MEASUREMENT

The history of measurement is almost as old as the history of man himself. Although we have no record of exactly where, when, and how the first measurement was actually made, we do know that man first had the need to measure about ten thousand years ago—about the same time that he became a settled, civilized being. Before that time, early man was a nomad, wandering from place to place as his food supply became exhausted. The caves and simple tents he lived in did not have to be measured. His food was gathered from what was available in his immediate surroundings, so he had no need to measure and plant fields. He supplied all of his own needs, so there was no need to be dependent on others and, hence, no need for measurements concerned with trade. However, once early man realized that he could raise his own food supply, he settled in one place. Because he was "farming," he had to measure fields and set boundaries. Because he was staying in one place, he had to measure and build houses of more permanent construction. Because job specializations led to the need for trade, he had to develop measurements to insure fair and equitable trading.

Faced with the need to measure, some chose the length of one of their feet as a unit. Others chose the number of grains of wheat or barley needed to measure desired weight. Still others chose different "units." There was no agreement on any one unit being used as a single standard by everyone. Even in ancient Greece and ancient Rome, where rulers generally decreed that their own feet would be the official units of length, problems developed because their feet were different lengths and before long, each city, state, and country had its own standard unit, different from all others. As cities, states, and countries began to have more contact with each other, chaos in measurement developed. By the late Middle Ages there was a complex assortment of measuring systems in use throughout Europe. Farmers, builders, traders, scientists, and all others who depended on measurement felt a need for the establishment of some kind of order.

THE METER IS INTRODUCED IN FRANCE

The first effort to introduce a single, logical, universal system of measurement came in the late eighteenth century after the French Revolution.

Scientific leaders in France, including Laplace, Lagrange, and Lavoisier, wanted to apply logic to every aspect of life, so they created a new measurement system with all units based on an unchanging natural measurement system, and with all arithmetic based on the number "10" for quick and easy calculation. The natural unit chosen for measuring length was one ten-millionth of the length of an imaginary line from the North Pole to the equator, passing through Paris. The unit would be called a "meter," which means "measure." All other units would be derived from the meter and the entire system would be called the "metric" system. French scientists constructed a platinum bar the length of a meter, which became known as the "standard meter."

To measure weight or mass, the natural unit chosen was the weight of water at 4 degrees Celsius, in a cube measuring one-tenth of a meter on all sides. The water temperature was specified because this is the temperature at which water is most dense, i.e., weighs the most. The French constructed a platinum cylinder 3.81 cm high and 3.81 cm in diameter as the standard unit of mass and called it the "standard kilogram," since all weight in the metric system would be measured in grams.

After the French had devised the metric system and had built these standards, they found that the line through Paris was 30.72 m longer than it was originally thought to be. The standard meter was therefore slightly shorter than it should have been. However, it was too late to change the entire system, so the platinum bar and not the line through Paris was accepted as the standard on which the system is based.

In 1799, the metric system became the legal system of weights and measures in France. The French people did not favorably regard the idea of converting from their existing system to the new one. Napoleon forced the system upon the people during his wars, but after that the system lapsed into disuse. By 1840, measurement in France, as in most of Europe, was in such utter chaos that the French government made it illegal to use any system of measurement other than the metric. Heavy fines were imposed for non-use of the metric system; and schools and businesses made concerted efforts to teach it. After 1840, the use of the metric system spread to most of Europe (not Great Britain) because it was so logical and easy to use. Scientists all over the world began using it as their official system and still do.

MEASUREMENT IN THE UNITED STATES

While France was developing the metric system, the United States was developing itself as a nation. One of the early concerns of the Founding Fathers was to establish a system of weights and measures. The Constitution gave this power to Congress. Congress placed Thomas Jefferson in charge of studying what would be best for the United States. He suggested that either a decimalized version of the English system or the French metric system could be used in the United States. Not knowing which of the two to choose, Congress decided to wait. In 1816, Madison reminded Congress that a system of weights and measures must be established. Congress charged John Quincy Adams with the responsibility of choosing a system. In his "Report upon Weights and Measures" (1821), Adams gave the pros and cons of both the English and metric systems. This document has since become a classic one in metric history. Adams also suggested that France, Spain, and Great Britain take a continuing look at the development of a common uniform standard. Because the United States still had close ties with Great Britain at the time of Adams's report, a brass duplicate of the English pound was brought to the United States in 1827, put in the Philadelphia Mint, and used as the standard for making coins. Soon, two bronze yardsticks were brought over. Thereafter, Congress passed laws making the English pound and yard the U.S. standards. Throughout the nineteenth century, the United States was developing into a world power. Feeling rather independent, the United States perceived no need for uniformity with other nations or within the United States itself. In 1866, Congress passed a law allowing the use of the metric system, but did not prescribe its use. The law read, "It shall be lawful throughout the United States of America to employ the weights and measures of the metric system, and no contract or dealing or pleading in any court, shall be deemed invalid or liable to objection because the weights and measures expressed or referred to therein are weights and measures of the metric system." The law applied to the weighing of gold, silver, and drugs ("Troy" weight) and to common things ("avoirdupois" weight). The English system had been improved and standardized by this time; thus there seemed to be no need for most Americans to change to metric, even though scientists preferred it.

THE LAST HUNDRED YEARS

In 1875, the International Bureau of Weights and Measures was established near Sèvres, Paris. New, more accurate copies of the standard meter and standard kilogram were made. Thirty-one exact copies of the standard meter were made and distributed to countries around the world. These bars were made of 90 percent platinum and 10 percent iridium so that they would not rust or change in length. They were made slightly longer in length than the actual meter, with engraved lines to mark off the exact meter length. In cross-section, the bars were in the shape of an "x," to prevent bending or sagging. Bars Number 21 and Number 27 were sent to the United States. Forty standard kilograms were also made and sent to countries around the world. The United States received weights Number 4 and Number 20. In 1893, meter bar Number 27 and kilogram weight Number 20 became the official standards for all length and weight measurements in the United States. They are kept in laboratories of the National Bureau of Standards in Gaithersburg, Maryland.

By 1900, 35 nations, including the major nations of continental Europe and most of South America, had officially accepted the metric system. By 1960, when the launching of Sputnik had generated a new interest in scientific ideas, almost the entire world had "gone metric." It was in 1960 that it was decided to abandon the "meter bar" as the international standard of length and to substitute a wavelength of light. The new definition was 1,650,763.73 wavelengths of the orange-red line produced by Krypton-86 = 1 meter. This new definition was a return to the original idea that the meter should be defined in terms of some unchanging natural standard. This new determination is accurate to 1 part in 100,000,000 and has the advantage of being reproducible in scientific laboratories all over the world. Also in 1960, *Le Système International d'Unités*, The International System of Units (abbreviated "SI"), was established. This was the official name for a standardized metric system to which all major countries agreed. It has seven fundamental units upon which all others are based: *meter*, *kilogram*, *second*, *ampere*, *kelvin*, *candela*, and *mole*. In 1965, Great Britain announced its intention to convert to the SI metric system. In light of this, the United States, in 1968, directed the Secretary of Commerce to undertake the U.S. Metric Study in order to determine the impact of metric conversion on America. In 1971 the study was reported to Congress. It recommended that the United States change to the International Metric System "deliberately and carefully" over a ten-year period.

A METRIC CHRONOLOGY

1788: The U.S. Constitution gave Congress the power to establish standard weights and measures.
The French Revolution stimulated thought; scientists wanted and needed a simple logical measurement system based on some unchanging natural unit and with an arithmetic based on 10. The meter and kilogram were developed as units of length and mass, respectively. Jefferson presented to Congress a plan for a unified measurement system, similar in structure to the metric.

1799: The metric system became the legal system of France.

1821: John Quincy Adams made a detailed report of the advantages and disadvantages of the English and metric systems.

1827: Congress adopted the English system, partially for political reasons.
Chaos reigned in French measurement; in 1840 a law was passed to make the use of any system other than the metric illegal. The metric system spread throughout Europe and most of the scientific community.

1866: Congress authorized the use of the metric system.

1875: The International Bureau of Weights and Measures was established.

1890–1900: Various departments of the U.S. government adopted the metric system for official use; 35 nations completely adopted the metric system.

1896: A bill requiring government adoption of the metric system was passed by the House of Representatives. Later, the House reconsidered its action and the bill was sent back to committee, where it eventually died.

1960: *Le Système International d'Unités* (SI) was established; all major nations except Britain and the United States adopted the metric system.

1965: Great Britain began metric conversion.

1968: The U.S. Metric Study Act was passed.

1971: U.S. Metric Study completed, and report made to Congress outlining a 10-year plan for conversion.

1975: A metric conversion bill was passed by the Congress and signed by the President.

REFERENCES

BENDICK, JEANNE. *How Much and How Many: The Story of Weights and Measures.* New York: McGraw-Hill, 1947. Grades 5–7. Pp. 160–67: chapter on metric system. Pp. 168–76: tables and illustrations about measurement, history of various systems.

FRISKEY, MARGARET. *About Measurement.* Chicago: Melmont Pubs., 1965. Based on Ford Motor Co. material, development of measuring units.

LUCE, MARNIE. *Measurement: How Much? How Many?* Minneapolis, Minn.: Lerner Pubs, 1969. Grades 3–6. Pp. 20–31. History of metric system and advantages of its use.

CHAPTER 2

How the Metric System Works

METRIC UNITS AND PREFIXES

In the SI system of measurement, the seven basic units are: *meter*, *kilogram*, *second*, *ampere*, *kelvin*, *mole*, and *candela*. However, since common usage centers more on length, volume, and weight, the development herein utilizes the *meter* (length), *liter* (volume), and *gram* (weight) approach for teaching the metric system. In particular, one may note that the liter is a non-SI term that came into usage before the metric system was modernized and that it actually denotes 1 cubic decimeter.

All metric units have a uniform scale of relation. The various units of measure get their names by adding prefixes to the units of meter, gram, liter, etc. Divisions of metric units are tenths, hundredths, thousandths, etc., and higher denominations are formed by multiplying the basic unit by 10, 100, 1000, etc. Metric numbers are therefore usually written decimally. The symbols (not abbreviations) m, L, and g denote *meter*, *liter*, and *gram*, respectively.

The table is taken from Appendix B and represents the more commonly used prefixes and their decimal equivalents.

METRIC RELATIONS

PREFIX	SYMBOL	MULTIPLICATION FACTOR	
mega	M	1 000 000	(millions)
kilo	k	1 000	(thousands)
hecto	h	100	(hundreds)
deka	da	10	(tens)
	m, g, L (basic unit)	1	(ones)
deci	d	0.1	(tenths)
centi	c	0.01	(hundredths)
milli	m	0.001	(thousandths)
micro	μ	0.000 001	(millionths)

EXAMPLES: A. prefix + basic unit = metric unit
centi + meter = centimeter
kilo + meter = kilometer

B. 1000 grams = 1 kilogram
1 meter = 100 centimeters
0.001 liter = 1 milliliter

Note in example B above, that the basic unit stays the same, while the prefix denotes the shift in position of the decimal. This is precisely the system we now use in our monetary system, e.g., 20 dimes equals 2 dollars, or 53 cents equals 0.53 dollars.

THE MOST COMMONLY USED UNITS

The metric system and the units of the system are related to one another by simple factors or multiples of 10 (powers of 10). It is therefore similar to the United States currency of dollars, dimes, cents, and mills. Most of us will need to learn only *meter*, *liter*, *gram*, and degree *Celsius*.

The metric units of *length* commonly used are:

millimeter	mm	for small dimensions
centimeter	cm	for daily practical use
meter	m	for large objects or short distances
kilometer	km	for long distances

The metric units of *weight* commonly used are:

milligram	mg	for small quantities
gram	g	for daily practical use
kilogram	kg	for large quantities

The metric units of *volume* commonly used are:

milliliter	mL	for small volumes of liquid
liter	L	most convenient measure of volume, and is preferred for liquid and dry measure
cubic centimeter	cm^3	for small volumes*

The terms *volume* and *capacity* are often used interchangeably. However, for the purpose of this book, the following distinction will be made. Volume will refer to the amount of space something occupies, while capacity will refer to the amount a container will hold. One generally uses different units of measurement when measuring volume and capacity. In particular, cm^3, m^3, and so on, are generally used to specify capacity; that is, the units used are those of dimension. For volumes, mL, L, and so on, are generally used.

* One cm^3 = 1 mL; the familiar convention *cc* will no longer be officially approved.

The metric units of *area* commonly used are:

square centimeter	cm^2	for small areas
square meter	m^2	for large areas

The metric unit of *temperature* is degree Celsius (°C). On the Celsius scale, pure water at standard atmospheric pressure freezes at 0° and boils at 100°. Thus, normal body temperature is about 37°, while a comfortable room temperature is about 22°. Appendix E contains a more complete temperature chart.

A chart, with pictorial representation of the above, has been produced by the Metric Information Office, National Bureau of Standards, and appears as Appendix F.

RECOMMENDED USAGE OF METRIC TERMS

In order to minimize problems in the use of the metric system by non-technical groups, including students through high school, the National Bureau of Standards has recommended the use of the spellings "meter" and "liter." In international technical communication, *metre* and *litre* may be used. Both spellings for both words are considered to be acceptable. All prefixes are accented on the first syllable.

For temperature, degree Celsius (°C) is used instead of degree centigrade. Also, *Celsius* is the only metric unit name that is capitalized.

Because the term "weight" is commonly used to mean "mass," the terms "weight" and "mass" are used as being synonymous in this publication. The National Bureau of Standards recommends that the use of "weight" for "mass" be avoided in technical publications.

Finally, plurals of symbols are never formed by adding a final *-s*. The symbol remains the same in the plural as in the singular, e.g., 10 kg, not 10 kgs.

EASY STEPS FOR METRIC CONVERSION

The figure provides a convenient and efficient device for computing decimal placement when converting within the metric system. It also points out the use of base 10 as an integral part of the metric system.

In the figure, you start with the prefix you have and count the number of steps it takes to reach the prefix to which you are converting. If you descend, you move the decimal point to the right the same number of places as steps counted. Similarly, if you step up to a larger prefix, move the decimal point to the left.

Thus, we have a very simple rule:

Larger—to the left. Smaller—to the right.

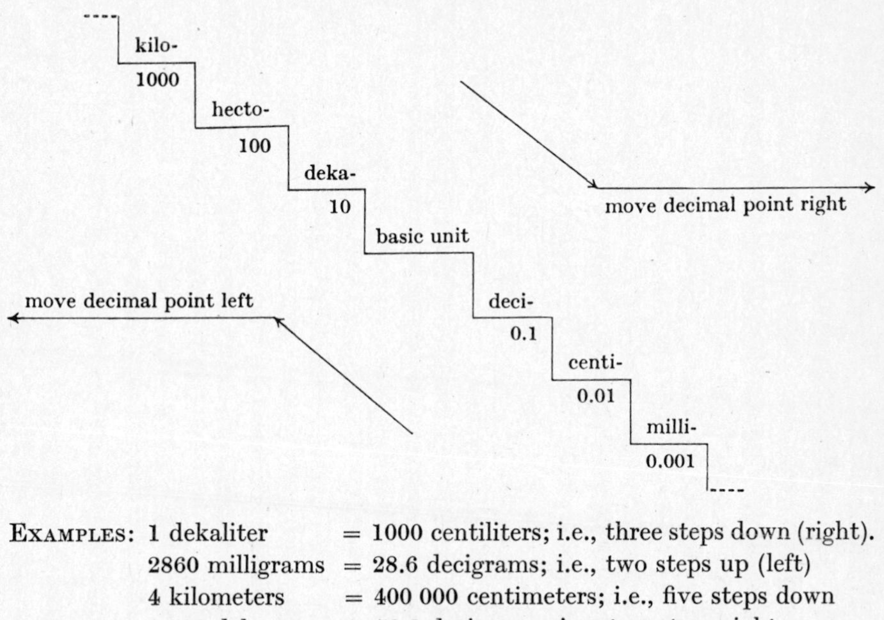

EXAMPLES: 1 dekaliter = 1000 centiliters; i.e., three steps down (right).
2860 milligrams = 28.6 decigrams; i.e., two steps up (left)
4 kilometers = 400 000 centimeters; i.e., five steps down
0.503 dekagrams = 50.3 decigrams; i.e., two steps right
= 0.005 03 kilogram; i.e., two steps left.

METRIC RELATIONSHIPS

The relationship between length, weight (mass), and volume (capacity) is as follows:

1 cubic decimeter of distilled water at 4 degrees Celsius at sea level has a weight of 1 kilogram and occupies 1 liter of space;

OR

1 cubic centimeter of distilled water at 4 degrees Celsius at sea level has a weight of 1 gram and occupies 1 milliliter of space.

Thus, the units of length, weight, and volume are all related.

Charts for supplementary units and conversion between the English and metric systems are contained in Appendixes D and E.

REFERENCES

EPSTEIN, SAMUEL. *The First Book of Measurement.* New York: Franklin Watts, 1960. Grades 4–6. Chapter on metric system, measuring temperature.

HOPKINS, ROBERT A. *The International Metric System and How It Works*, 2nd ed. Tarzana, Calif.: Polymetric Services Inc., 1974. The most complete source of SI information available; a veritable encyclopedia of the metric system.

TUSSELL, SOLVEIG PAULSON. *Size, Distance, Weight: A First Look at Measuring.* New York: Henry Walck, Inc., 1968. Grades 4–6. Pp. 37–45: metric system defined in simple terms.

CHAPTER 3

Teaching the Metric System

METRIC READINESS

It is apparent that the metric system is being adopted almost universally, but the United States is the only major industrial nation that has not made the transition. With all evidence pointing toward the use of this system, it is imperative that our educational institutions should not delay the process of teaching the metric system of measurement to every student and should be first in creating an atmosphere of "thinking metric" in the United States.

The real secret to learning the metric system easily and quickly is to learn to "think metric." It is the idea of "thinking metric" that is most important. The immediate response from most adults and students is to approach the metric system by trying to memorize lists of numerical conversions and equivalences between two systems. Real difficulties in learning the metric system will occur if people try to convert everything back to the American system. Clearly, there will be a need to know a few estimations, e.g., a meter is a little longer than a yard, or a kilogram is a little heavier than 2 pounds. But more than this tends to confuse. For the most part, people will not need to convert between the two systems.

Persons experienced in "think metric" procedures suggest that children will be able to make the transition fairly easily, while older people are likely to be thrown badly by the distortion of familiar dimensions. Mr. Kenneth Fetter, Director of Curriculum Services in the Division of Curriculum and Instruction at the New Jersey State Department of Education, suggests the following procedures that can be accomplished prior to implementation of metric education at the local level:

1. Select a committee or individual to be responsible for being the metric authority in the school.
2. Inventory all instructional equipment and materials. This inventory should be kept current and available in a central location.
3. Reorganize instructional media so that metric units of measure can be introduced in the primary grades.
4. Provide the instructional staff with in-service programs that stress involving children with concrete measuring activities.
5. Encourage the entire staff to become members of an organization that will keep them current as to information and teaching techniques.

The main educational advantage of the metric system lies in the simplification of teaching and learning how to measure. This advantage arises from the simple interrelations of units based on multiplication and division by powers of 10 and from the ease of computing with decimal fractions and whole numbers. Teachers are thus able to concentrate on a system that is logical, efficient, and more easily understood. The time saved in teaching this simpler system can then be used for the introduction of other materials. Some of the customary drill in fractions can be reduced, owing to the decimalization of the system. Also students will not have to learn a new system of measurement for science classes.

Since subunits and multiples in the metric system are related by powers of 10, the student must understand the decimal numeration system, including decimal fractions, in order to change from one related unit to another. Also, the teacher should help the students understand the vocabulary of the metric system. Thus, the teacher should provide readiness experiences for learning the metric system, as well as considering those children of differing abilities to understand decimal fractions.

Some of the curriculum changes that will be brought about by metrication include—

1. An earlier introduction of decimal concepts, with corresponding reinforcement of the place value system.
2. A downplay of skills in manipulations of fractions.
3. A general upgrading in the teaching of measurement.

Teachers should also view the metric system with some perspective, i.e., once the United States has converted to the metric system, and the school texts reflect this fact, teachers will probably spend no more time on the "meter-liter-gram" system than they now spend on the "foot-pound-quart" system.

METRIC CONTENT FOR K–8

Primary: K–3

1. Exposure to the metric system.
2. Idea of relation (small vs. large, hot vs. cold).
3. What basic units measure:
   How long, how wide, how tall, how high = meter.
   How heavy = gram.
   How much will it hold = liter.
4. Work with the following units:

centimeter	gram
meter	kilogram
kilometer	liter

5. Identify and use a meter stick.
6. Identify the following equivalents:
   100 cm = 1 m 1000 g = 1 kg 1000 m = 1 km
7. Use a 30 cm ruler for measurement.

Grades 4–5

1. Background: the early primary content.
2. Pronounce names associated with the metric system.
3. Recognize the symbols used for unit prefixes in the metric system.
4. Work with addition and subtraction of metric quantities.
5. Identify °C.; boiling and freezing points of water.
6. Become familiar with the prefixes:
   deci deka
   centi hecto
   milli kilo
7. Identify the following equivalents:
   1000 milli (units) = 1 of same unit
   100 centi (units) = 1 of same unit
   10 deci (units) = 1 of same unit
   1 deka (unit) = 10 of same unit
   1 hecto (unit) = 100 of same unit
   1 kilo (unit) = 1000 of same unit

Grades 6–8

1. Extend use of basic units.
2. Extend operations and calculations within the metric system.
3. Expose to larger and smaller units—find areas where these are used.
4. Emphasize decimal operations.
5. Expose to scientific notation—negative exponents.
6. Relate to other areas (money, physics, exponents).
7. Applications for daily use.

SAMPLE TEACHING UNIT : SEVENTH GRADE

I. Introduction
   A. History of measurement
      1. Logic of early units
      2. Fallacies of individualized standards
   B. Reasons to "Go Metric"
      1. Need for uniform standard
      2. Similarity to monetary and numeration systems

II. The Metric System
  A. Units of measurement
    1. Linear—meter; weight—kilogram; volume—liter
    2. Temperature—degree Celsius
  B. Use of Prefixes
    1. deka, hecto, kilo, and larger units
    2. deci, centi, milli, and smaller units

III. Metric Exercises
  A. Review of decimals and their manipulation
  B. Conversion of smaller units to larger units
  C. Conversion of larger units to smaller units

IV. Metric Activities
  A. Linear metric measurement
    1. Uses of meter, decimeter, centimeter, millimeter
    2. Estimation of measurement
  B. Metric area measurement: measurement projects, geoboards, etc.
  C. Metric mass measurement
    1. Use of kilogram and other units
    2. Use of metric balance scale
  D. Metric volume measurement
    1. Use of liter, centiliter, milliliter, other units
    2. Graduated cylinders and finding volumes

V. Supplemental Activities
  A. Twenty questions, metric quiz, "metro," oral drills
  B. Flashcards, activities

METRICS FOR YOUR COLLEAGUES AND THE COMMUNITY

Why Use the Metric System?

As of January 1, 1977, there were only four countries in the world not committed to the metric system: Brunei, Burma, Liberia, and Yemen.

While people in the United States today—even though metric conversion legislation has been enacted—are not required to use the metric system, they are using it more and more each year. In the near future it will probably be the only system of measurement used in the United States and the rest of the world.

Some underlying reasons for adopting the metric system in the United States are:

1. The metric system is simpler than the English system, with everything arranged in tens and the promise of saving much time and confusion.
2. The sciences, such as chemistry, physics, and biology, already rely on the metric system.
3. Engineering, manufacturing, and world trade are making use of the metric system more and more all the time.

The Effect on Business

Some U.S. corporations several years ago started to convert to the metric system. The two main reasons for metric conversion in industry are international trade economics and the simplicity of working with the metric system. Ford Motor Company and General Motors have already begun a metrication program, and as a result, machine tool sales have increased. International Harvester and Caterpillar Tractor have metric programs that require all new designs to be in metric units. IBM and the Timken Company will completely convert to the metric system because their export trade has been suffering. The Ohio Department of Transportation has already installed road signs that show distances in kilometers as well as in miles. Highway departments in Arizona, Michigan, and Minnesota have begun similar metrication programs. The California Wine Institute has proposed that domestic winemakers convert to the metric system.

A list of other companies already using the metric system in some way include General Electric, Litton Industries, Northrup Corporation, Gerber, Nabisco, Xerox, Honeywell, North American Rockwell, McCormick, and Sears, Roebuck. Also note the following:

1. Sears, Roebuck and Company's sales of tools, which feature both metric and English-unit products.
2. The statutory standard for automobile emissions was defined in metric units.
3. Metric units at many sporting events and the metric dimensions at Cincinnati's baseball stadium.
4. The dual labeling of food products and dress patterns.
5. The metric units on medical prescriptions.

The Effect on the Consumer

The indirect effect of metrication on the consumer will be considerable.

1. Some of these will be the costs of retooling.
2. The costs of repackaging.
3. The costs of new weighing equipment for fresh foods.
4. Learning new packaging names.

Since business must make a profit, these costs will be passed on to the consumer. However, the process of conversion will also offer many opportunities to eliminate superfluous varieties in sizes of products, parts, and containers.

The American National Standards Institute took no position on whether the United States should or should not go metric, but ANSI intends to be prepared. The American Home Economics Association is responsible for the administration of a very important American National Standards Committee that is going to be deeply involved with the metric system as the United States converts. The committee has produced an American National Standard that is a basic document for those working with recipe development.

Gardening

Not too long from now, American gardeners will find themselves measuring garden rows in meters instead of feet and inches; spray solutions as so many milliliters per liter; and temperatures in degrees Celsius instead of degrees Fahrenheit. Seeds will be packeted by the gram instead of the ounce; and fertilizer will probably come by the kilogram, if not the metric ton. Sizes of nursery plants will be described in meters and centimeters.

Farming

Blossoms on our fruit trees will freeze at 0° instead of 32°, greenhouses will be kept in the vicinity of 20°, and if they are still pasteurizing soil at that time, agronomists will subject it to heat above 100 degrees. Instead of acres, people will be thinking in terms of hectares.

Foods

The impact of metrication on the consumer will be both direct and indirect. Some juggling will have to be done in the marketplace when purchasing a liter of milk, a half kilogram of butter or margarine, or perhaps, butter and margarine will be put up in kilogram packages, the equivalent of about two pounds. But in this packaged world, the tendency is to measure the package visually rather than examine the label for content of the package.

Clothing

For the home sewers, many fabrics and sewing notions are currently imported and are manufactured to metric measurements but are then converted to the inch-yard system for the American market. In the near future, such products will simply retain their metric measurements.

Furniture

In purchasing furniture and household textiles, buyers get out the measuring tape or yardstick to make sure the purchase is the appropriate size or amount. However, except for rugs and carpeting, shoppers generally think in terms of inches, not in feet and yards. It should not be difficult to convert to centimeters once a metric ruler, a metric stick, and a metric tape become available.

Monetary System

The United States is one step ahead of some foreign countries that have recently gone metric. Since the U.S. money system is already decimalized, the nation can maintain its present system for figuring finances.

IDEAS FOR CONDUCTING A METRIC WORKSHOP

I. Introduction
  A. A brief history of measurement
  B. Short history of the metric system
  C. Why study the metric system?
II. How the Metric System Works
  A. Our monetary system
  B. The arithmetic of the metric system
  C. Meter—liter—gram
  D. Metric prefixes and their relationships
  E. Experiences and activities using metric devices
III. What to Teach and When
  A. Kindergarten–3
  B. Grade levels 4–5 and 6–8
  C. A continuous program through the grades
IV. Open discussion and Evaluation to Implement Your Ideas
  A. Prepare brief packets containing information, handouts, etc.
  B. Have the participants actively involved in metric activities; include the construction of something metric, estimating, and measuring
  C. Show films, discuss texts, give demonstrations, build an exhibit, etc.
  D. Put the "Think Metric" idea into every item mentioned above

REFERENCES

The Arithmetic Teacher, Washington, D.C.,: The National Council of Teachers of Mathematics, Inc. The April 1973 issue is devoted entirely to metrication articles.

JACOBSON, WILLARD J., CECILIA J. LAUBY, and RICHARD D. KONICEK. *Inquiring into Science*, vol. 5. New York: American Book Co., 1965. Pp. 249–54: grade 5. Description of metric system, plus four pages of activities.

SRIVASTAVA, JANE JONAS. *Weighing and Balancing*. New York: Thomas Y. Crowell, 1970. Grades K–3. Demonstrates different ways to measure items. Can be used for metric system measurements.

CHAPTER 4

Metric Activities

This chapter is devoted to a multitude of activities and suggestions for making the metric system easy and fun to learn. Most of the activities have been used at a variety of student levels. As metrication is relatively new to all levels, "lower levels" will herein denote primary grades K–3, while "upper levels" will denote 4–8. However, since there are students in the lower grades who can handle more advanced concepts, and conversely, since there are students in the upper levels who have difficulties with basic measurement, the activities in this chapter are presented in a general sense within the upper- and lower-level categories.

This thus allows teachers to adapt the various activities to the needs of their students and to their particular teaching situation. Experiences in many workshops and in-service institutes have shown that specifying "Activity A" for students of "Level B" is not as effective or stimulating for the teacher as saying, "Here is an activity, or idea. In how many ways, or settings, can you use it?"

LOWER GRADE LEVELS

Bulletin Boards

Measuring Match

Make a bulletin board containing pictures on one side and measurements on the other. For the teacher's convenience, the following three bulletin boards have been correctly matched. Each should be reordered, and the students asked to match them correctly.

WEIGHTS		TEMPERATURE		VOLUME	
(pictures)	*(wgts.)*	*(pictures)*	*(temp.)*	*(pictures)*	*(vol.)*
penny	3 g	skier	−10 °C	milk carton	1 L
mouse	70 g	people	37 °C	test tube	15 mL
baby	4 kg	ice	0 °C	measuring cup	250 mL
football player	100 kg	boiling	100 °C		
kitten	250 g	summer	40 °C	gas tank	50 L

"Guess Which Is Larger"

Make a bulletin board showing optical illusions containing line segments. Discuss which segments are longer; measure to verify. Change illusions every few days.

Sayings

Familiarize children with "old" sayings that will change with metrication, e.g., "Walk a country kilometer." Use these sayings on a display or bulletin board.

They Like to Compare

LENGTHS	HEIGHTS	WEIGHTS
Bulletin board	Themselves	Marble
Toys	Doors	Bottle cap
Arm reach	Desks	Buttons
Shoe	Chair	Coins
Shoestring	Light switch	Chalk

Using Metrics to Measure Height and Weight and Length

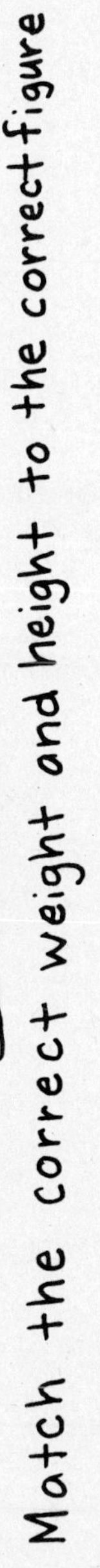

Reference Key

1 meter = 1.1 yds
1 centimeter = 0.4 inch
1 Kilogram = 2.2 lbs

Decimal Bulletin Board

Put a seven-digit number on the board and have a child move the decimal point to the left or right. The child calls on someone to read the new number. If he reads it right, he then comes to the board and moves the decimal point to a position he wants. He calls on someone then.

Matching Bulletin Board

Place on bulletin board or chalkboard the following words:

liter *meter* *gram* *centimeter* *kilometer* *kilogram*

Have children collect pictures and place them under the word that could be used to measure the article in the picture.

OR

Mix up the pictures brought in and have the children rearrange them under the correct measurement that would be used.

Games

Capture a Pair

Draw line segments measuring different lengths on thirty 3″ x 5″ cards so that three of each length are drawn in three different colors. (There will be a total of ten different lengths.)

Deal three cards to each player and put the remainder of the deck in the center as the "kitty" face down on the table. The player to the left of the dealer starts; he draws a card from the kitty; if he has two matching segments he puts them face up on the table. If another player has a matching segment he can, in turn, "capture a pair" and claim all three cards.

A player may challenge another's attempt to capture a pair. If the challenge is correct, he captures the pair and buries them in the kitty. (Use a metric ruler to verify challenges.)

Play until one player runs out of cards or until the kitty is empty. To score, add 2 points for each set of three, 1 point for each pair of two and −1 point for each single card left in a player's hand.

Metro

Play this game like Bingo, but use cards with metric abbreviations instead of numbers and with "METRO" written across the top instead of "BINGO." The child who wins calls "Metro" and becomes the caller for the next game.

Note: The tabs used by the caller should contain the name of the unit; the markings on the "Metro" cards should be the symbols (see samples).

This game could be adapted to other concepts within the metric system that the teacher wishes to emphasize, such as metric equivalents, prefixes, etc.

M	E	T	R	O
dg	dam	m	kL	mg
hm	dm	dag	daL	kg
mL	L	cL	m	dL
hL	hg	mm	cm	km
g	cm	kg	cg	hm

Sample "Metro" Card

R
"centimeter"

Sample Callers Tab

Metric Checkers

Prepare a set of questions concerning metrics (if 3 x 5-inch cards are used, answers could be put on the backs). Play checkers in the usual way except: Before a player can move his checker he must answer a metric question correctly; if he cannot answer correctly, he may not move on that turn.

Variation: Prepare your own checkerboard by drawing a square of desired size on tagboard; divide the square into 64 smaller squares; in alternate squares in each row (like the black squares on a regular board), write a question concerning metrics. Before a player can move to a square he must answer its question correctly.

Twenty Questions

The teacher selects one object and describes it by saying, "I am thinking of something that is ____________." For example, it might be an eight-centimeter-long chalk. Each student is allowed to ask one question or to make a guess that requires a yes or no. Whoever guesses right is the new leader. First describe the object by its dimensions, surface area, volume, or weight.

Unscramble the Measurement Words

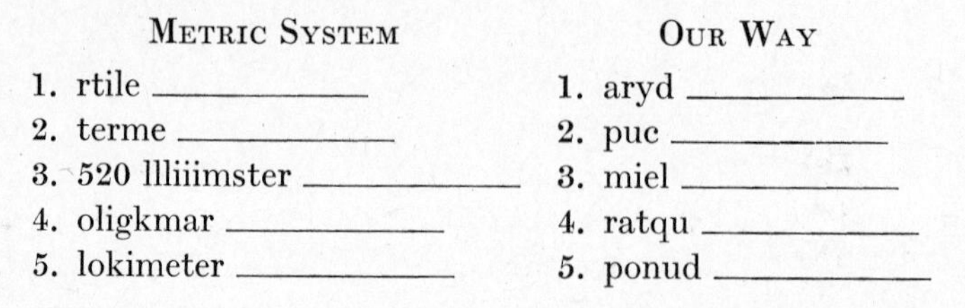

METRIC SYSTEM	OUR WAY
1. rtile ____________	1. aryd ____________
2. terme ____________	2. puc ____________
3. 520 llliiimster ____________	3. miel ____________
4. oligkmar ____________	4. ratqu ____________
5. lokimeter ____________	5. ponud ____________

Metric Hangman

Play hangman on a blackboard using words and phrases from the metric system. Use the same rules as in regular hangman. Children guess what letters are in the word. Each wrong letter adds to the man in the "noose."

"Now You See It, Now You Don't"

Scramble six basic metric prefixes and the words *meter*, *liter*, and *gram*. While students put their heads down, take one away. With heads up, which word is missing?

Shortcut

This game requires a basic game board with "shortcut" (see sample) and four sets of cards, each set of a different color. The red cards contain easy metric questions, the blue average questions, the green difficult questions, and the brown (shortcut cards) questions of great difficulty. Provide a spinner with red, blue, and green areas and let the players take turns spinning it. If the pointer lands on red, the player must answer a question from the red pile. If he answers correctly he advances his marker one space; if incorrectly he goes back one space (except from "start"). If the spinner lands on blue, the player can choose a question from either the blue pile or the red pile. Blue indicates a move of two spaces. If the spinner lands on green, the player can choose from any of the three piles; it indicates a move of three spaces. The pile of brown "shortcut" cards can be chosen only when a player lands on a "shortcut" square. If he answers correctly he may take the shortcut; if he misses, he must go back to the beginning and start again. When a player lands in the shortcut square he has the option of picking a card from any of the piles. If he picks from any of the other piles, he may not take the shortcut. First player to reach "finish" wins.

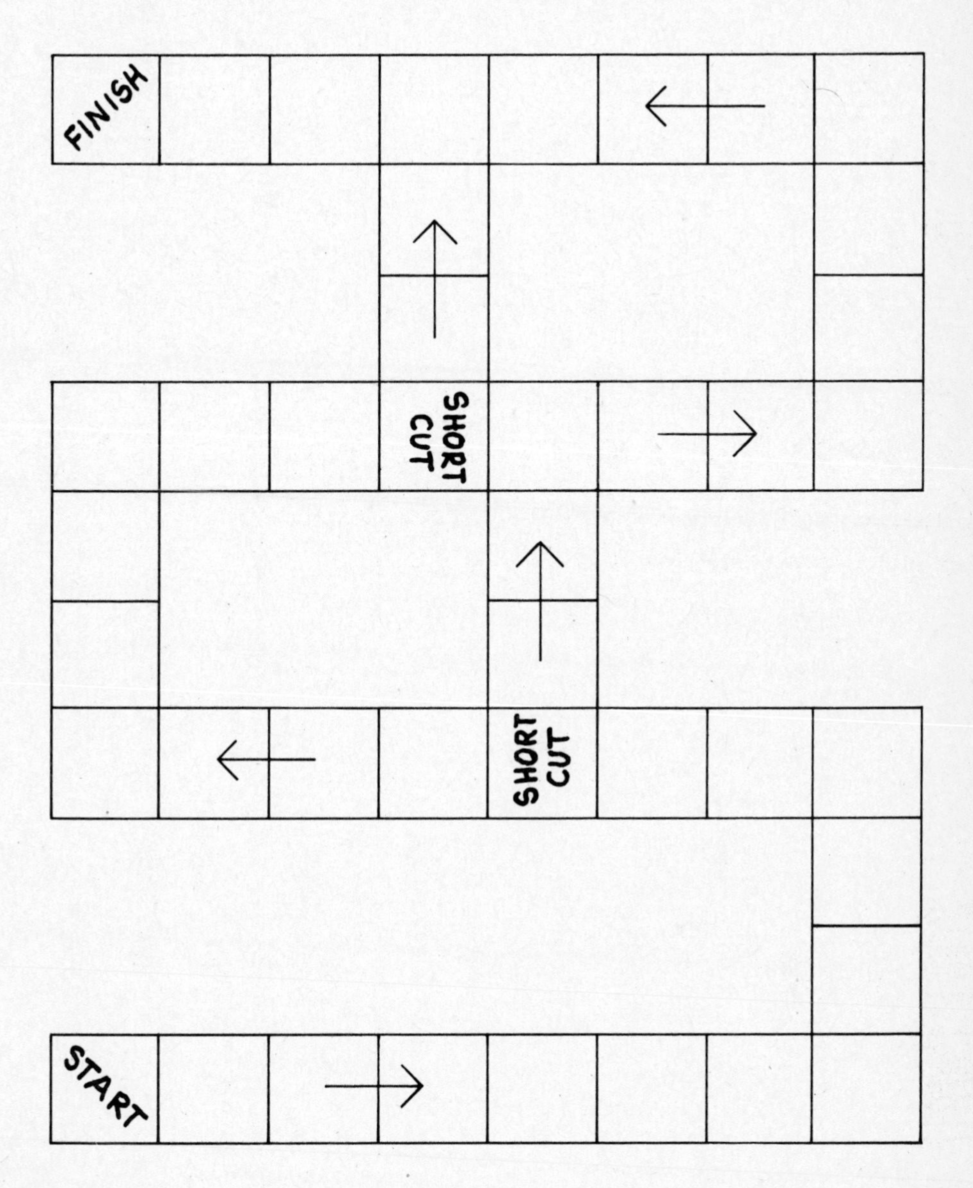
FINISH
SHORT CUT
SHORT CUT
START

Match Game

Students are asked to match prefixes with symbols, meanings, or equivalent amounts. Some sample cards (squares) follow:

cm	cg	hg	$\frac{1}{100}$
kg	10	g	100
$\frac{1}{1000}$	m	1000	L
mL	$\frac{1}{10}$	dm	dag

Card Games

Metric Deck

Materials: 25 blank cards or index cards
3 different colored pens

Make up a deck of "metric cards." This deck has three suits, each a different color and unit:

meter (length); *liter* (volume); *gram* (weight).

Each suit contains eight cards, i.e., one for each of the following prefixes combined with the basic unit.

micro—10^{-6} μ	deka— 10^{1} da
milli— 10^{-3} m	hecto—10^{2} h
centi— 10^{-2} c	kilo— 10^{3} k
deci— 10^{-1} d	mega—10^{6} M

The joker is a card with English units of measure written over the face of it.

Sample cards:

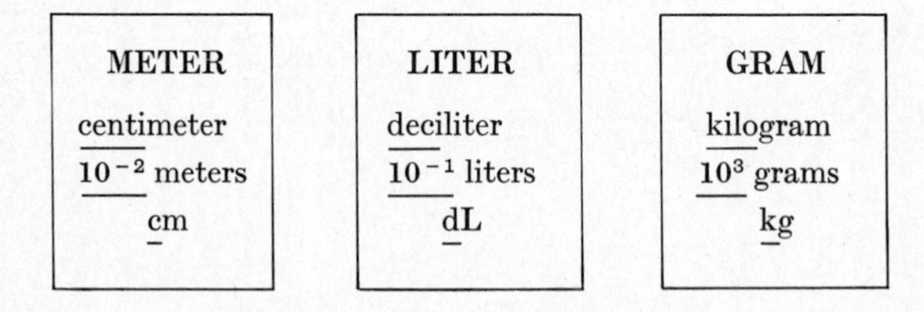

Shuffles

The dealer shuffles the deck and passes four cards to each player and then places the remaining cards face down beside him. When everyone is ready, the dealer draws a card from the top of the deck and either replaces it with one from his hand or passes it face down to the next player. The cards are passed rapidly in this manner. The winner is the first one with a set of three cards with the same prefix.

"Metric Moe"

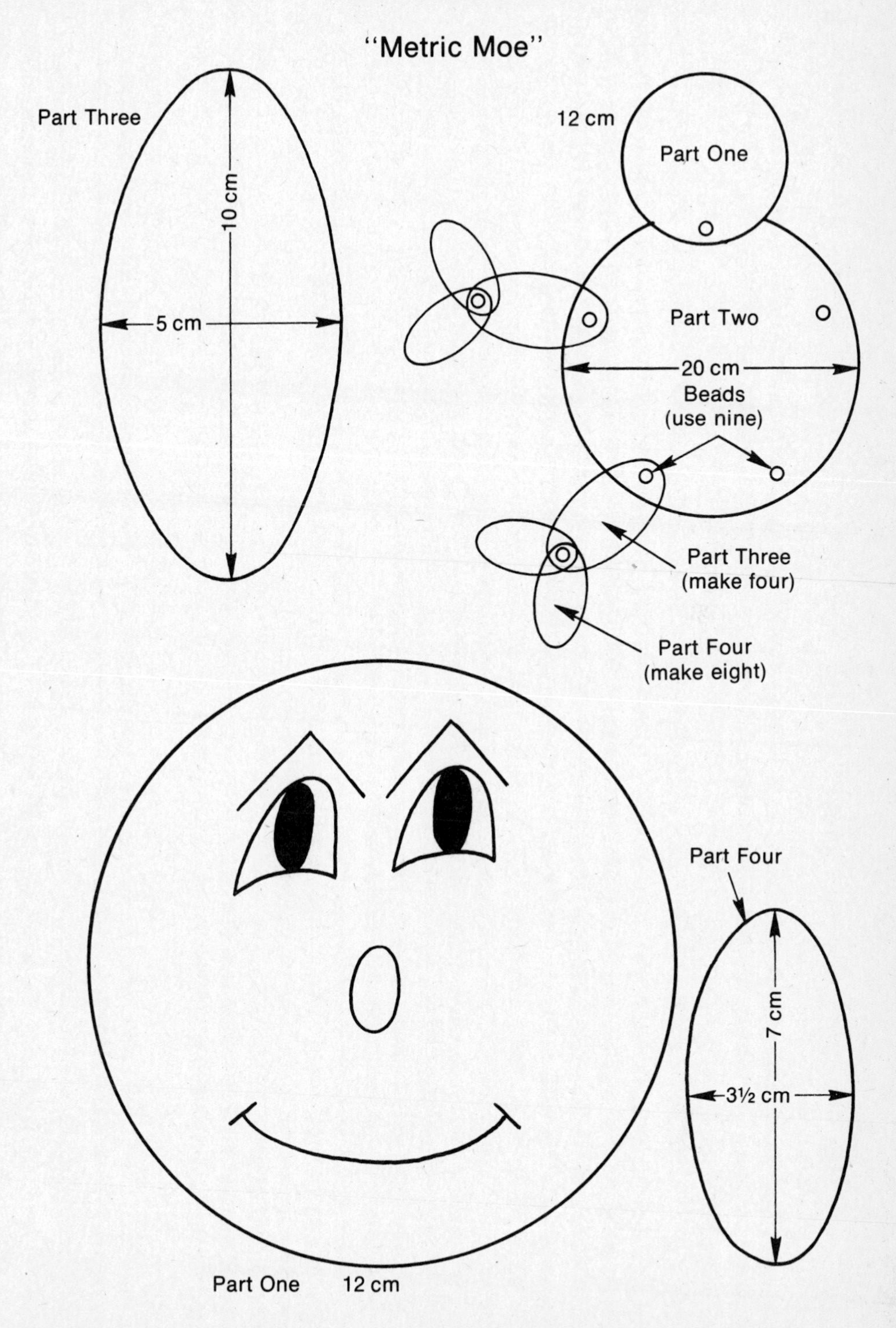
Part Three
10 cm
5 cm
12 cm
Part One
Part Two
20 cm
Beads
(use nine)
Part Three
(make four)
Part Four
(make eight)
Part Four
7 cm
3½ cm
Part One
12 cm

Skunk

Add the joker card so that the deck now has one "odd" card without a pair. Deal the 25 cards. Each player in turn draws a card from the player on his left. When the player has a match, he places the pair on the table. The player who is left with the joker is the "skunk."

Metric Rummy

Deal five cards to each of three players. Each player may draw from face-down deck or face-up discard. Players lay down sets of three or more cards that either:

1. Have the same prefix, for example, centimeter, centiliter, centigram.
2. Are consecutive in the same unit, for example, milligram, decigram, centigram.

Players may play on opponent's runs. Points are scored one per card. The first to go out gets a 10-point bonus. Others subtract points in hand from laid-out cards.

Go Fish

Deal four cards to each player. Stack remaining cards face down in the center of the table. Players take turns drawing one card off the top of the deck. If the card does not match prefixes on any card in the player's hand, it is discarded face down at the bottom of the deck. If it matches, the pair is placed on the table, and the player takes another turn. The first player to match all his cards is the winner.

Measuring Activities

Measuring Volume

Mix one liter of tempera; use two primary colors to make another color to spark interest. Use this for painting that week. Pour what is left over into a liter container for comparison with the original amount.

Allow a liter of juice per table of five or six children, and have a juice and cookies "party."

Have a child bring an empty container from home. Use water to see if it holds a liter. With supervision, allow a small group to pour and compare. (Best done outside.)

UPPER GRADE LEVELS

Class Projects

New Advertising

Have children collect and bring in ads in the old English system of measuring and as a class convert each approximately to metric. Have students make new labels and use items in a metric supermarket set up in one corner of room. Children can play in "supermarket" in spare time and can practice using metric in reference to food weights and measures.

Aim: To become familiar with size and contents (how much).
Construct and operate a store:
Which size container would be best for a small or large family?
What size would be best for one person?
Would boxes of different shapes contain the same amount?
(Teacher could make up her own questions).

Temperature

Aim: To become familiar with boiling point, very hot days, comfortable days, winter readings, and freezing point.

Make a thermometer with cardboard and red ribbon and mark the points:

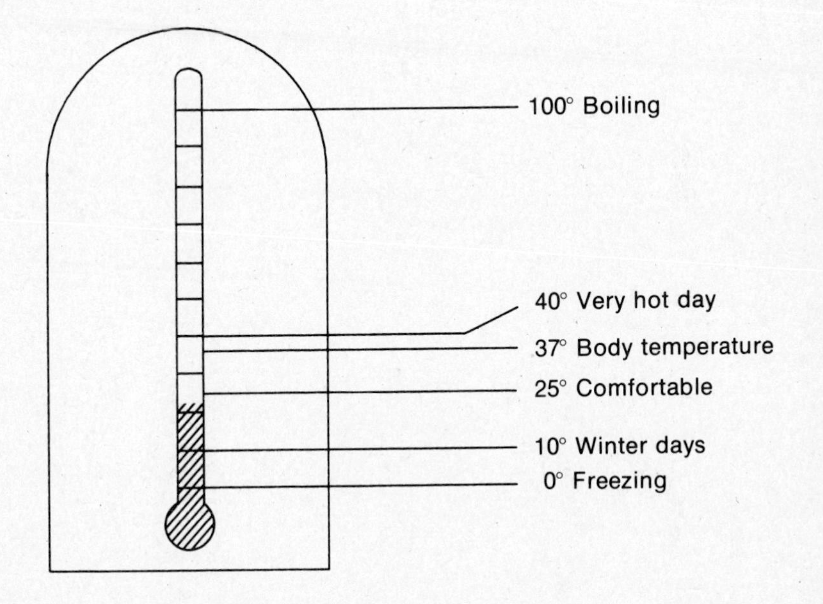

Electricity

Make a chart for readings of electric meter (dates—meter index—kilowatts used). Have students read meters once a week for a month and calculate the number of kilowatts used during the month. Ask the students to look at their parents' electric bills, noting how many kilowatt hours were used. Calculate how much energy was used by the class. Research on the electric meter: Assign a short paper in which the students describe what they think it would be like if we were thrown into the metric system tomorrow.

Metric Ecology

Measure rainfall in a metric rain gauge.
Measure tree heights in the metric system.
Measure the rate of flow of a nearby stream.
Measure the temperature each day and compile a record.

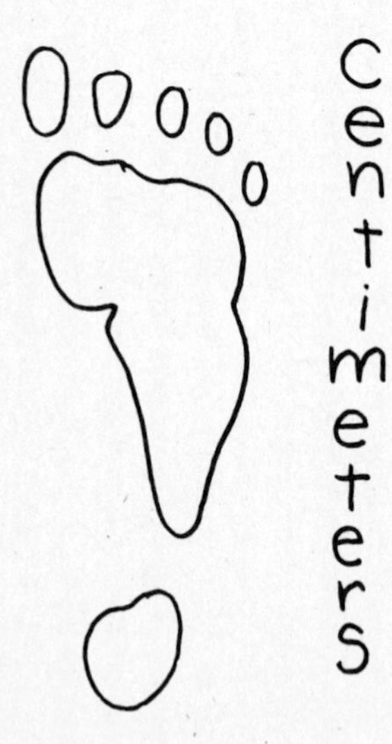

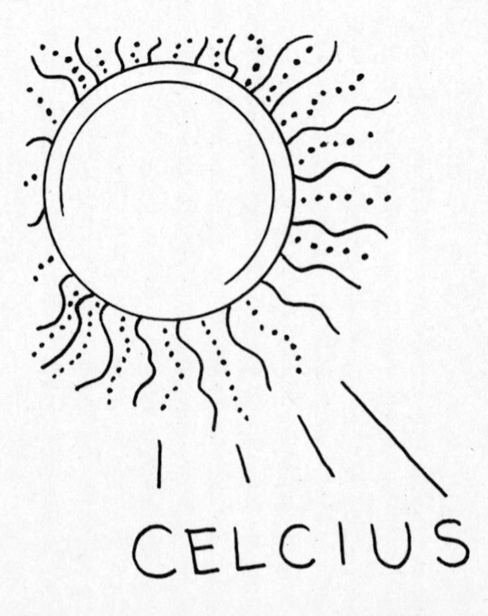

Metric Computer

Decorate a box to look like a computer. Cut one slot for INPUT and one for OUTPUT. One student sits behind the box with paper and pencils to be the BRAINS of the computer. Other students write a metric measurement problem on a slip of paper and put it into the INPUT slot. The computer computes and turns out the answer through the OUTPUT slot. Anyone catching an error from the computer exchanges places with him.

Plan a Trip

Plan a trip with a service station roadmap. Make up metric legend 1 cm = _____ km.

Using a thread, measure distances along highways and then find distances on a metric ruler.

Find the cost of gasoline per liter; speed limits; time of trip; kilometers per liter. Find cost of trip, distances, etc.

A Trip to Europe

Hello, my name is ___________. I am in the second grade. My Mom and Dad are taking a trip to Europe, and they are going to take me. I just got a new camera and a back-pack.

My Dad has a lot of books about Europe. He tells me many things are just the same as the things we have in the United States. He also tells me many things are different in Europe.

When our airplane lands in Paris, France, I will not be able to understand the people because they speak French. It will be the same way in Italy, Spain, and Greece. But when we get to England, I'll understand the people because they speak English.

Another thing that is different in Europe is the way they measure things. When I weigh myself there I will be weighed in ___________. In our school the nurse weighed me. I weighed _____ pounds. In Europe I will weigh _____ kilograms. Eventually, everyone in America will be weighed in kilograms because it's easier.

In the past, my mother sent me to the store to buy two pounds of peaches. To get the same number of peaches in Europe, I will get about one kilogram of peaches. You can see that one kilogram is about the same as ___________.

When I measure myself in Europe, I will be measured in centimeters instead of inches. In Europe I'll be _____ centimeters tall.

My Dad is going to buy a car in Italy. He tells me that when we buy gasoline there we will not buy gallons of gas, but ___________ of gas. The gas gauge and the speedometer will already be marked in the metric system for us.

When we look at a road sign to see how far we are from Paris, it will not be told in miles but in ___________. The speedometer on our new car will not be in miles per hour, but in kilometers per hour.

If we need milk for our meals in Europe, we''ll buy a __________ of milk instead of a quart of milk. A liter is just a little bigger than a quart.

I am really looking forward to my trip. Many things will remind me of home, but many things will be new and different.

In Europe I'll say:
kilometers not miles
kilograms not pounds
meters not yards
centimeters not inches
a liter of milk
a kilogram of peaches
50 kilometers per hour
15 kilometers to the picnic
Soon, in America, we'll all be saying and using the metric system.

Measuring Activities

Mobiles

Make mobiles or suspend from a long string or colorful yarn different cutouts with the metric measurement printed on them. Use same idea for equivalence in measurements.

Metric Rulers

Using sturdy oak tag or balsa wood have each child make his own ruler for metric measurement. A good size would be about 20 cm. Tell students that the size of the stick they have is 2 dm. Each will divide his in half to mark decimeters. When he has found that to measure accurately he must have small divisions, help him to divide each dm into 10 smaller units (centimeters). These rulers can be used all year in measuring.

Catch the Error

Prepare stories containing many references to measurement. Include several errors. For example: "The dog grabbed the 10-gram cat and carried it over the 50-centimeter stream." Students learn to estimate and notice errors. The first one to find an error earns a point. The highest score wins.

Drawing Figures

Give children a number of 3 cm x 3 cm squares. The teacher holds up a larger premeasured rectangle or square with sides multiples of 3 (6 cm x 9 cm, 3 cm x 12 cm, etc.) and asks the students to arrange their squares to make the same shape.

As children make the shape, the teacher goes around and places the shape she is holding on top of one the child made; comparisons of "smaller," "larger," "the same," etc., can then be made. The teacher should have about 20 rectangles or squares ready to hold up.

Build a Metric Train

The children have centimeter sticks from 1 cm to 1 dm (they should have several of each). The teacher holds up a card containing a length and asks the students to build a train of that length.

Note: Cuisenaire rods may be used.

Sample Teacher cards:

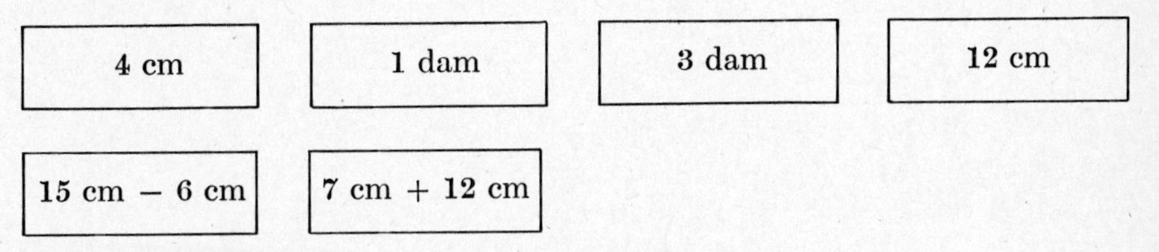

Metric Tanagrams

The following figure is a 9 cm square. Thus it can also be used for length, area, and conversion exercises, as well as the game of tanagrams.

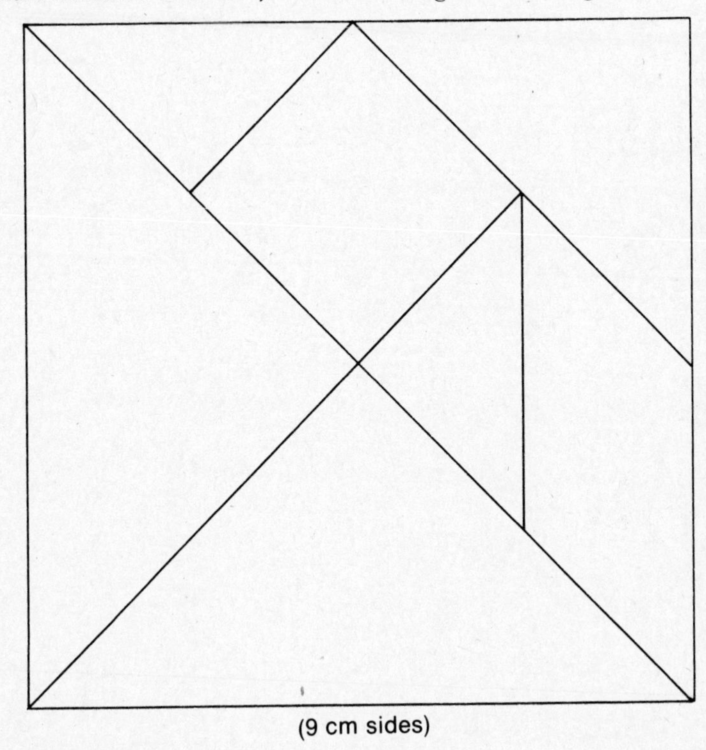

(9 cm sides)

"What's Your Area?"

Rule an actual-size, rectangular centimeter grid like the one shown in reduced size.

1. Take off a shoe. Put your foot on the centimeter grid. Trace around your foot. Count the number of square centimeters in your own footprint. How do you count the parts of the squares? Area ____________. What is your shoe size (length)? ____________.
2. Now place your hand on the centimeter grid and trace around it. How wide is your hand? ____________. How long is your hand? ____________. What is the area of your hand?

Graphs

The grid represents 1 cm squares. Plot the following points on the graph by locating them first on the *a* (horizontal) scale and then on the *b* (vertical) scale. For example, the point (9 cm, 20 mm) of Number 1 would be located at a = 9 and b = 2, since 20 mm = 2 cm. Number the points as you plot them and connect them with straight line segments in numerical order.

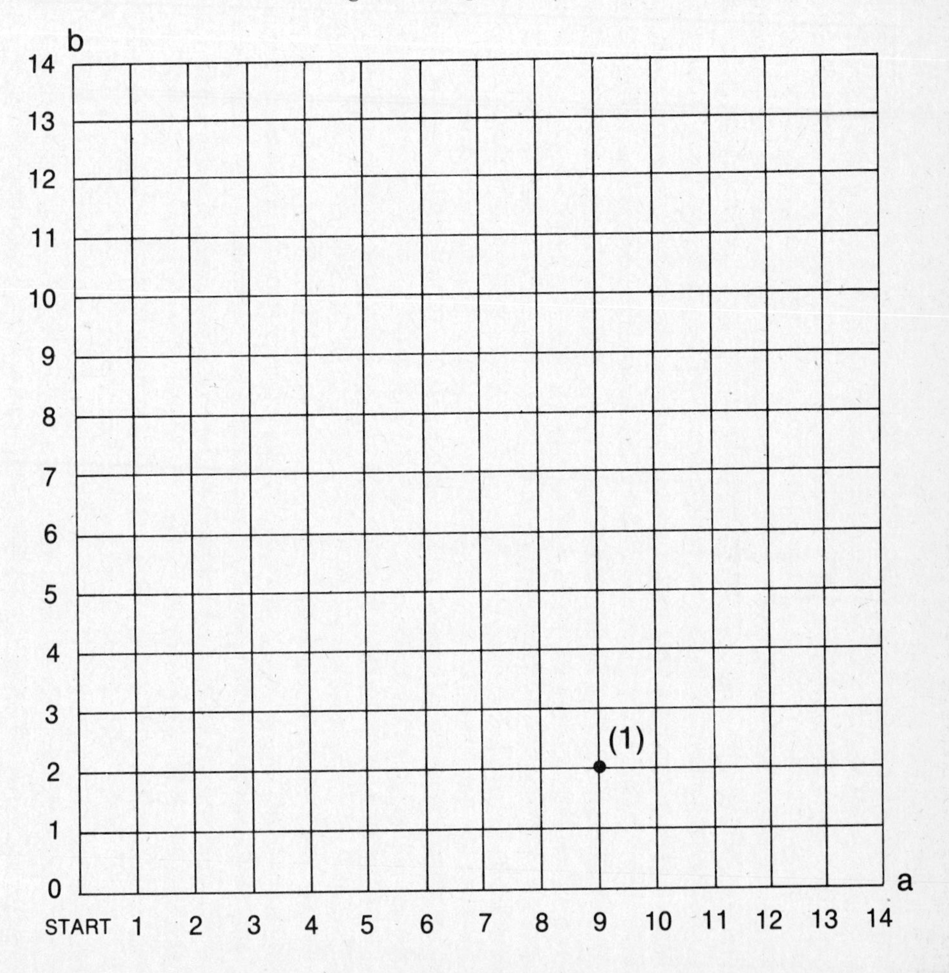

1. (9 cm, 20 mm)
2. (105 mm, 4 cm)
3. (0.13 m, 0.4 dm)
4. (115 mm, 6.5 cm)
5. (1.3 dm, 0.09 m)
6. (10.5 cm, 0.9 dm)
7. (90 mm, 0.11 m)
8. (7.5 cm, 9 cm)
9. (0.5 dm, 90 mm)
10. (65 mm, 0.65 dm)
11. (0.05 m, 0.04 m)
12. (7.5 cm, 40 mm)

Dominoes

Cut out the squares. Fit them together so that the touching edges name the same lengths.

<table>
<tr><td>0.057 mm
0.4 cm 0.01 mm
0.023 km</td><td>23 m
4 cm 10 dm
57 cm</td><td>230 m
400 km 40 hm
57 dm</td><td>23 mm
40 m 400 m
5.7 km</td></tr>
<tr><td>570 mm
10 cm 0.4 mm
230 dam</td><td>57 hm
400 hm 10 dam
0.057 cm</td><td>57 km
0.01 m 4 m
23 dam</td><td>230 cm
0.1 km 4 dm
23 cm</td></tr>
<tr><td>23 hm
0.04 dam 0.04 mm
570 km</td><td>570 m
4 km 10 m
2.3 cm</td><td>5.7 m
0.01 mm 400 dm
57 mm</td><td>2.3 dm
1 mm 4 mm
57 m</td></tr>
<tr><td>23 km
40 dam 1 dm
2300 mm</td><td>0.023 hm
40 dm 0.1 cm
5.7 hm</td><td>5.7 cm
0.1 mm 40 km
230 km</td><td>570 dm
1 dam 40 mm
230 hm</td></tr>
</table>

Games

Metric Grab Bag

The class makes up metric problems. Assign each a number and list the answer on the answer key alongside the number. Each student should assign a point value of 1 to 5 points. Place the problem slips in a bag. Each student draws out one problem. Points are gained for correctly answering a problem and lost for errors. Highest score wins.

Metric Bee

A spelling bee using metric measurement questions.

Baseball

Make up four sets of metric questions so that each set is a little more difficult than the one before it. Label the sets from the easiest to the hardest: "single," "double," "triple," and "home run."

Draw a baseball diamond on the board or set one up within the room, using the corners of the room.

Divide the children into teams. Play the game following the same rules as regular baseball, except that each child coming to bat chooses what he wants to hit. The teacher then asks a question from the appropriate set and the child either advances to the proper base or is out. The teacher should act as umpire and "rule" on each answer.

Concentration

Purpose: To encourage students to become more rapidly familiar with metric measures.

Method: Sample arrays are provided below for a "Concentration" game, using the overhead projector. Other arrays may easily be made up on your own. After making transparencies of the arrays, they should be placed on an overhead projector and the measurements covered with pieces of cardboard or coins. The class should be divided into two teams. Then, taking turns, each team tries to uncover pairs of equivalent measures by calling out the letter and number of a position in the array, for example, "A3 and C2." The teacher then uncovers these positions to see if there is a match. Relative to the first sample array below, a guess of A1 and D2 would be a match, since 100 cm = 1 m. In the second sample below, A3 and D3 provide a match, since g stands for gram. Points are awarded for a match, and that team retains the "turn." If a guess is wrong, the positions are re-covered, and the other team takes a turn.

	1	2	3	4
A	100 cm	4 L	1 kg	2 m + 5 m
B	25 g	3 cm + 9 cm	7 m	12 cm
C	1 L	5 kg	17 g + 8 g	1000 g
D	10 L − 6 L	1 m	1000 mL	18 kg − 13 kg

	1	2	3
A	m	centimeter	gram
B	cm	milliliter	kg
C	liter	mL	meter
D	L	kilogram	g

Variations of this game can be played to help the students learn equivalents. The game can be extended to involve mass and volume equivalents.

Tic-Tac-Toe

Purpose: To help students to more readily recognize equivalent metric measures.

Method: Have students prepare a table of equivalent metric measures, extending from the kilometer to the millimeter, for a game in linear measurement; from the kilogram to the milligram for mass measurement; from the kiloliter to the milliliter for volume measurement. The children may refer to the table as needed to play the game or this may be excluded as students' facility with equivalents increases.

Metric Tic-Tac-Toe uses three dice (either from a math operation math kit or handmade). The first die contains on its faces: 1 cm, 1 mm, 1 dm, 1 m, 1 km 1 hm. The second die contains arithmetic operations +, +, −, ×, ×, ÷. The third die contains 1, 100, 1000, 10, 10 000, 0.1. Variations of the above to include mass and volume may be worked out. Dice are shaken in a shaker cup and spilled thus: 1 cm × 1000 = ______. On a Tic-Tac-Toe board each player would insert his answer and attempt to fill a winning three in a row with the same number answers or any requirements that may be made up.

Make up Tic-Tac-Toe charts similar to the ones illustrated. Use the chalkboard, poster paper, or overhead projector if the whole class is to be involved. Charts may be made up as worksheets to allow for group work or working in pairs.

Winners are determined as in regular Tic-Tac-Toe.

100 m	1 m	100 cm
100 dam	10 dam	10 m
1000 m	10 cm	1 hm

10 mg	1 kg	1 g
1 dg	10 cg	100 mg
100 g	10 hg	1 dag

10 dL	1 kL	1000 mL
10 hL	100 cL	100 daL
1 L	10 hL	100 dL

The example may be varied to improve students' use and understanding of metric prefixes. The first die should contain prefixes *kilo*, *hecto*, *deka*, *centi*, *deci*, and *milli*. The second die may be changed to include only × and ÷, i.e., three × signs and three ÷ signs. Use a shaker cup and devise a means of scoring for a group or class teams.

Metric Recipes

Converting Recipes to the Metric System

Bring in a number of recipes from cookbooks or collections of recipe cards and rewrite the recipes using metric units. Remember that 1 fluid ounce is equivalent to approximately 34 milliliters, and a pound is about the same as

454 grams, i.e., 1 kilogram is approximately $2\frac{1}{4}$ pounds. Use an unmarked cup similar to a kitchen measuring cup and mark it according to metric units.

By measuring the amount of fluid in each, work out a set of milliliter equivalents for:

$\frac{1}{2}$ teaspoon	2 tablespoons
one teaspoon	4 tablespoons
one tablespoon	

New England Pot Pie (The Hartford Courant)

- $1\frac{1}{2}$ kilograms cooked turkey in large dice (3 pounds)
- 125 grams onion, chopped (1 small onion)
- 125 grams carrot, cubed (1 medium carrot)
- 75 grams celery, chopped (1 medium celery stalk)
- 5 grams salt (1 tsp.)
- 60 centigrams white pepper ($\frac{1}{4}$ tsp.)
- 45 grams butter (3 tbsp.)
- 125 grams bacon, diced (4 slices)
- 40 grams flour (4 tbsp.)
- 20 grams corn relish (1 tbsp.)
- 1 pre-baked pie crust

Cover cooked diced turkey, onion, carrot, and celery with 1 liter (1 quart, 2 ounces) boiling water. Add salt and pepper, cover, and simmer 45 minutes or until vegetables are just tender. Drain broth, reserving 300 milliliters ($1\frac{1}{2}$ cups) for sauce. Melt butter and sauté bacon until lightly browned. Stir in flour. Cook, stirring constantly, about four minutes. Pour reserved broth into flour mixture and cook until thickened. Stir in corn relish and combine with turkey mixture in a deep casserole dish. Cover top with crust and bake in a 190 °C (375 °F) oven 20 minutes, or until warmed through and golden brown on top. Serves four to six.

Veal Paprika

- 700 g veal, cubed
- 60 mL chopped onion
- 120 mL strained tomatoes
- 60 mL flour
- 5 mL salt
- 45 mL fat
- 1 mL pepper
- 120 mL water
- 120 mL sour cream

Mix flour, salt and pepper, and meat. Sauté meat, onion, and paprika to taste (until browned). Sieve tomatoes. Add water and meat, and simmer 90 minutes. Add sour cream, and simmer 15 minutes more. Serves four.

Hawaiian Punch

- 1430 mL vanilla ice cream
- 720 mL pineapple juice
- 75 mL orange juice
- 15 mL lemon juice

Beat ingredients. Serve in chilled glasses. Yields 3 liters.

Metric Salads Are Fun!
1/2 liter mushrooms
2 onions
2 sprigs parsley
1 head lettuce
10 milliliters salad oil
OPTIONAL:
1/4 liter green pepper
1/4 liter carrots
5 milliliters wine vinegar
Combine
Add salt, pepper,
toss and
Serve!

Kilogram Cake

Grease and line with paper a 24 x 12 x 8–centimeter loaf pan. (It replaces the old 9 x 5 x 3-inch pan.)

1. Sift together into a bowl:
   500 g flour — 10 g baking powder
   250 g sugar — 5 g salt
2. Add:
   125 g soft shortening — 5 egg yolks
   5 mL vanilla — 110 mL milk
3. Beat two minutes then add:
   55 mL milk

Mix together. Spoon batter into prepared pan. Bake 60 to 70 minutes in a moderate oven, 185° (Celsius). Cool and ice with orange glaze.

Puzzles

Metric Crossword Puzzle Number 1

Across

2. Measures how hot it is.
4. Measures how much water.
6. °C is the symbol for degree ____________.

Down

1. Measure things in 10.
3. Measures how far.
5. Measures how much you weigh.
7. Measures how much time.

Metric Crossword Puzzle Number 2

The outline around the puzzle may be an animal, toy, or a seasonal item. You can easily add and delete portions to suit your needs.

Across

1. Milk is bought by the ___________.
2. A paper clip weighs one ___________.
3. The ___________ system is based on 10.
4. A paper clip is about _____ centimeters long.
5. A pack of gum is _____ centimeters long.

Down

1. A meter measures ___________.
2. The length of my ___________ is less than a meter.
3. My father is two ___________ tall.
4. A pen is about _____ centimeters long.

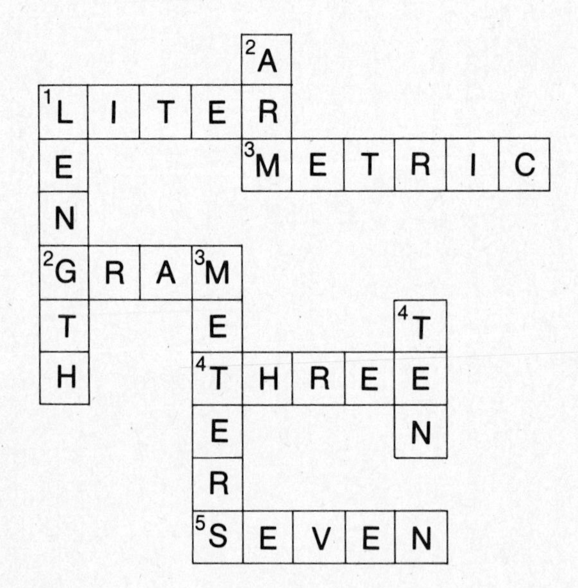

Physical Activities

Treasure Hunt

Make up a list of items. Teams of students are to collect as many as possible in 20 minutes. The winning team is the one with the most correct items.

Suggestions: a red ball that weighs between 10 and 100 grams, a green leaf of area less than 36 square cm, a can label that lists between 400 and 500 grams, a bottle of fewer than 250 milliliters.

Dashes

Measure and run the 50-yard and 50-meter dash.

Relay Race—(2 teams). One by one children go to blackboard, pick a metric problem from a box and solve; if wrong, next teammate tries. First team solving five problems correctly is the winner.

Football

The field is either mapped out metrically by the students or by the teacher (for young children). The length of the field will vary according to grade level, e.g., for the third grade, the field might have the dimensions of 50 meters by 25 meters.

Ten down-markers are marked off to play. Normal touch football rules are followed (four downs to score). Students are required to figure out how far they must go to make a touchdown after each down.

Standing Broad Jump

Have each child jump and then cut a string the length of his jump. Compare farthest, shortest, same distance with other children or have the children measure jumps with meter sticks and compare lengths.

Enrichment Topics

A Maximum Mental Metric Meditation

The paper industry can benefit by stocking only the maximum size paper most frequently used. Smaller sizes can be obtained by repeated halving of the sheets until the desired size is obtained. All sheets of paper from the newspaper to the postage stamp can be similar in shape. Determine the size of a "metric newspaper." It is to be rectangular in shape and one square meter in area. When it is cut in half, the shape obtained is similar to the original shape. Prove that repeated halving produces similar shapes.

"Metrics, Metrics, Everywhere!"

History: Look up Jean Delambre and Pierre Méchain in reference books to find the exciting story of their Dunkirk-Barcelona survey, done under wartime conditions. Role-play the history of measurement and the metric system.

Radio: Look up the prefixes used in the metric system and show other places where they are used and their meaning in various contexts. Radio broadcasting: megahertz, kilohertz.

Science: Trip into the science room—investigate the various metric equipment available.

Medicine: If possible, get a doctor or pharmacist to speak on "The Importance of the Metric System to Me."

Mechanics: Find a "Metric Tool Set" (many people have these for their cars or motorcycles). Compare sizes of other tools. Find out which cars, motorcycles, etc., use these tools. Why are they easier to use?

Small Photography: What is an 8 mm movie projector? Have you ever heard of a 35 mm cartridge for your camera? Students interested in photography can investigate it and the metric system.

Small Photos: Take pictures of the areas where metric is now in use; make a bulletin board of these photos. Also, make a display of these to help motivate the community to go more metric (could be displayed in some community building).

Painting: Measure the walls or floors of the classroom, home, school, etc. Find the areas involved and estimate the number of liters of paint needed to paint the respective areas.

How many more can you find?

Estimation and Matching

This is a self-test. Draw lines to connect each object and its measurement.

I. Linear Matching	
	A. 1 km
1. Baby	B. 55 cm
2. First-grade child	C. 10 m
3. Your teacher	D. 160 cm
4. Center on basketball team	E. 500 cm
5. Football field length	F. 1 m
	G. 210 cm
	H. 91 m

II. Weight Matching	
6. Baby	A. 1000 kg
7. First-grade child	B. 50 kg
8. Your teacher	C. 250 g
9. Football player	D. 3.5 kg
10. Car	E. 23 kg
	F. 105 kg
	G. 5000 g

III. Volume Matching

11. Aspirin tablet	A. 100 mL
12. Thimbleful	B. 41 mL
13. Cup of soup	C. 2 kL
14. Soda bottle	D. 4 mL
15. Tank of gasoline	E. 0.1 mL
	F. 60 L
	G. 250 mL

Metric Riddles

Make up or have students make up metric riddles.

Metric Rhymes

Design a contest with awards for the best metric rhymes. Some examples follow.

Let's have some appreciation
in these days of tribulation
for rules of metrication.

Length is measured by meter
and volume is measured by liter

Hams, lambs, and yams
can all be weighed with grams.

Choose a book that has a map
Place the book upon your lap
Consider distance, oh gentle readers
And convert the miles to kilometers.

Nation after nation has made the swing to metric
Isn't it time we took the plunge
And made measurement less hectic?

ALL GRADE LEVELS

Measuring Activities

Why not try—

Investigation of historical units of measure.
Growth of plants and animals.
Estimating and checking capacities of common containers.
Making graphs of capacities and volumes of containers.
Weight gain of classroom pets.
Noting weight gains of students during the school year.
Scale drawings involving house floor plans, rooms.
Making a kite with a tail one meter long.

You can also—

A. Place a metric stick on a table for a small group and ask them to see if they can roll a clay "snake" one meter long. Compare results on metric stick. Again, use the terms centimeter and millimeter as questions arise.
B. Using a science or astronomy book, obtain the distances between the sun and the planets in miles. Compute the distances in kilometers.
C. Measure and compare circumferences and diameters of cylindrical objects in metric units to determine the value of pi.
D. Name as many advantages as you can that the metric system has over other systems of measurement.
E. Using some type of balance scale, compare weights of objects and of metric weights (if metric weights aren't available, a one-inch paper clip = 1 gram, and a nickel = about 5 grams).

Lengths

Use a piece of thread to follow along the following curves. Then stretch out the thread and measure it, using a handmade or commercial metric ruler.

Find the lengths of the line segments involved in each pattern or figure below and then find the lengths.

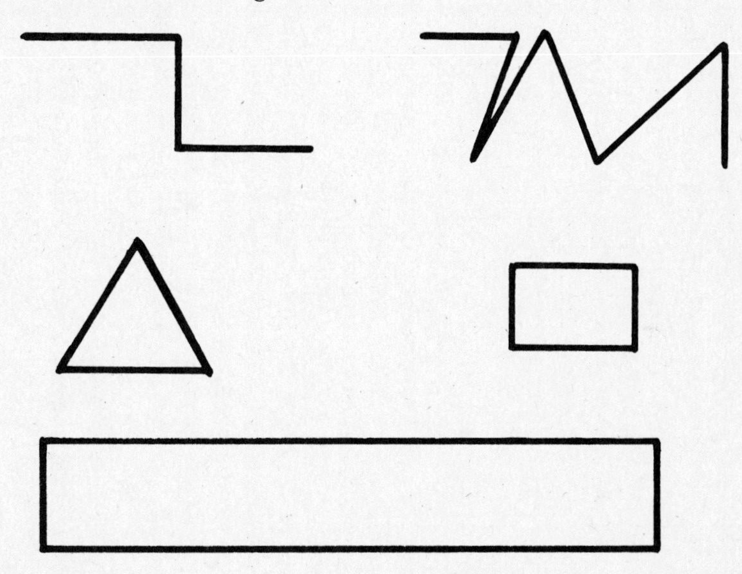

Metric Athletics

This is more-or-less a group project to be presented to the class. Each group finds a sport that would be altered if metrication were achieved in the United

States. *Examples:* Boxers' weight classes would be in kg and gloves in gm. Football gridiron in meters. Running distances in meters. Weight lifting in kg.

As a culmination, play a game or sport activity that has been converted to the metric system (everyone does the 50-meter dash).

Simon Says

Children hold up a pencil point for a mm, a finger for a cm, and their hands spread for the measurement of a meter.

Miscellaneous

Make individual metric tapes; divide these into dm, cm, and show the tiny size of the mm. Use the metric tapes to measure designated objects in the room.

Give metric measurements and have children find items with these measurements.

Tape meter strips on the wall and have students measure their own heights in meters and centimeters.

Make a meter trundle wheel—each time the wheel goes around 1 meter is measured.

Measure circles, wheels, and tires with string marked in cm and m.

Have students measure their bicycle tires in meters; mark the tires, then count the number of times the tires goes around by watching the marks. In this manner have students measure the distances from home to school or other distances.

Frizbee Throw

A grid is mapped out by the students or the teacher (for younger children). Markers are placed 5 meters apart. The total length is 30 meters. Students throw frizbee for distance. For upper grades, accuracy may also be scored; e.g., the number of meters to the left or the right of the straight line is subtracted from the distance.

Volume

Using the following pattern, construct a cubic decimeter. Could you visualize a cubic millimeter and cubic meter?

Use the cubic decimeter to help you complete the following sentences.

For lower grades:

A. The cubic centimeter fits in the cubic decimeter ______ times.

B. The cubic centimeter fits in the cubic meter ______ times.

C. The cubic decimeter fits in the cubic meter ______ times.

For upper grades:

D. The cubic millimeter fits in the cubic centimeter ______ times.
E. The cubic millimeter fits in the cubic decimeter ______ times.
F. The cubic millimeter fits in the cubic meter ______ times.

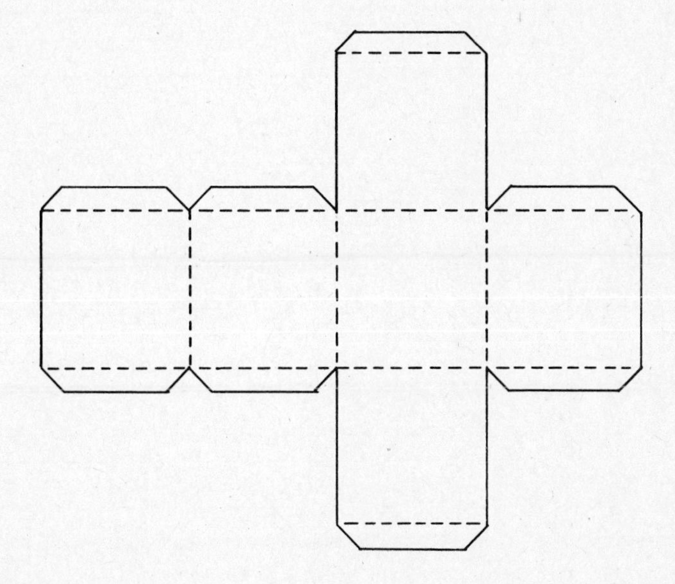

Test Your E.Q.

E.Q. = Estimating Quotient!

Estimate the amount of water contained in an aquarium, the length of a rod or dowel, the weight of a book, the size of a sheet of paper, etc.

Have a ballot box in the classroom and a tear-off pad of paper to use for ballots. Have children estimate and cast ballots. Closest estimate wins. (Can have a prize such as a metric ruler or blue ribbon, or can have the child receive the "honor.")

Change the item to be estimated once every week or two.

Prepare a fish tank: first weigh everything separately; then weigh finished tank.

Geoboards

Make geoboards, in 1 cm units, and use in explaining material or supplementary work with areas of shapes in the metric system. Children can use them on their own and discover other properties of area or cm^2 or relationships in the metric system.

"How Would You Measure?"

Using familiar objects, have the class decide what metric unit each should be measured in. This can be strictly length or you can combine volume and weights as units to measure with.

Metric Number Line

Have class member (or group of children) construct a number line, labeling it with metric units of length. These number lines can then be used to measure and compare objects. Also, since number lines are useful in basic operations, the metric number line can be applied to problems using the metric system.

Measure the U.S.A. in a Metric Way

Materials: Metric ruler; string; scissors.

To answer these questions use the accompanying map, string, and ruler. This type exercise can be made better if a local map or a larger more colorful map is used.

1. The average width of Texas is ______ kilometers.
2. The average length of Illinois is ______ kilometers.
3. The area of Colorado is about ______ square kilometers. This is ______ square meters, or ______ acres. An acre is approximately 4000 square meters. Using the correct prefix, this is ____________.
4. What is the largest state shown on the map? ____________
5. Its area is approximately ______ square kilometers.
6. What is the largest of the Great Lakes? ____________
7. Find the distances in metric units between the following:
   a. Augusta, Maine, and Lincoln, Nebraska ______.
   b. Boston, Massachusetts, and Gallup, New Mexico ______.
   c. Miami, Florida, and Seattle, Washington ______.
   d. Detroit, Michigan, and Chicago, Illinois ______.
   e. Little Rock, Arkansas, and Las Vegas, Nevada ______.
   f. St. Paul, Minnesota, and Billings, Montana ______.
   g. Alantic City, New Jersey, and San Francisco, California ______.

Span

Using their hands as a unit of measure, students can be asked to take the measurement of objects in and around the classroom in "spans." Children will see the need for a standard measure as they find that objects being measured vary in length because everyone's hand (span) is different.

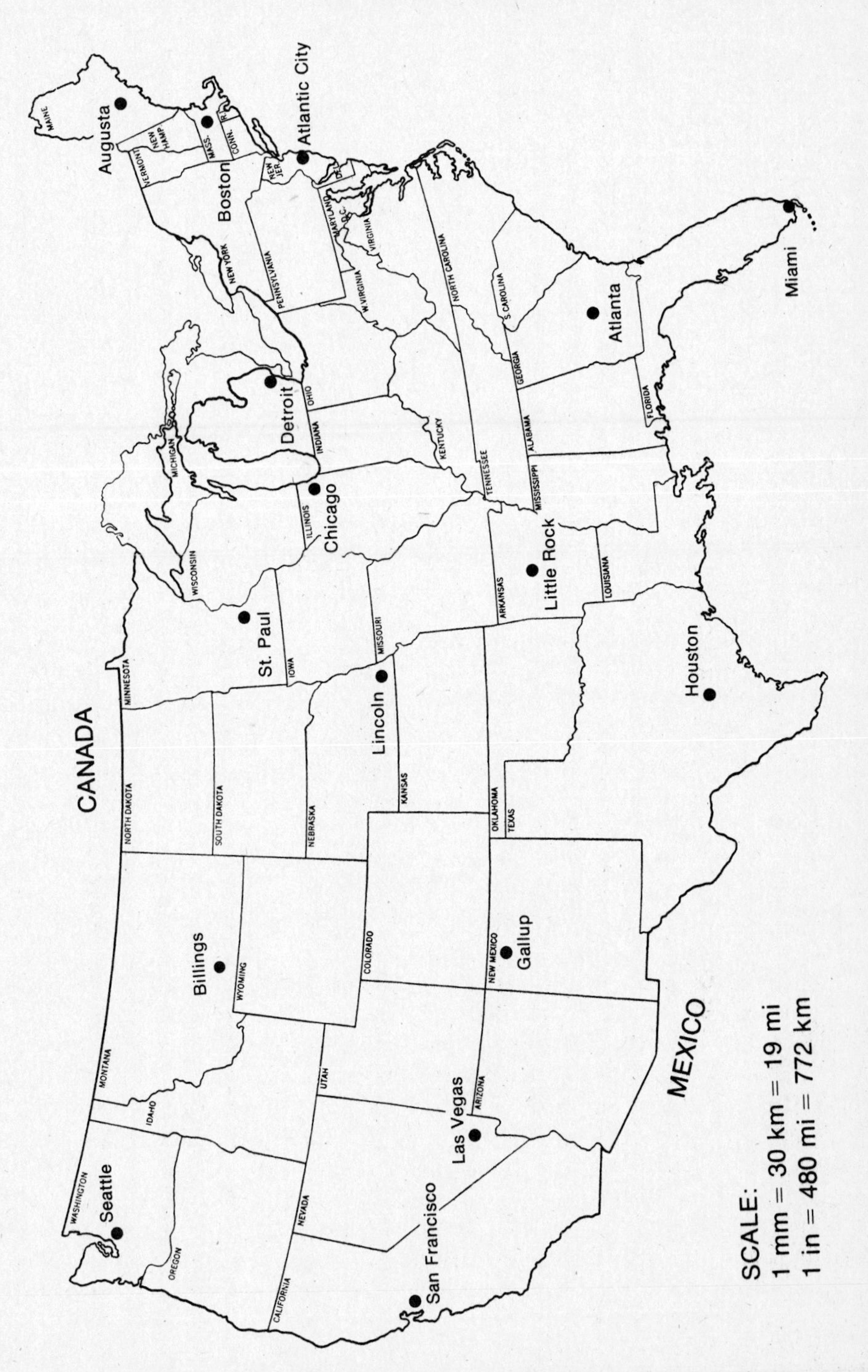
CANADA
MEXICO
Seattle
San Francisco
Las Vegas
Billings
Gallup
Lincoln
St. Paul
Chicago
Detroit
Little Rock
Houston
Atlanta
Miami
Boston
Augusta
Atlantic City
WASHINGTON
OREGON
CALIFORNIA
NEVADA
IDAHO
MONTANA
UTAH
ARIZONA
WYOMING
COLORADO
NEW MEXICO
NORTH DAKOTA
SOUTH DAKOTA
NEBRASKA
KANSAS
OKLAHOMA
TEXAS
MINNESOTA
IOWA
MISSOURI
ARKANSAS
LOUISIANA
WISCONSIN
ILLINOIS
MICHIGAN
INDIANA
OHIO
KENTUCKY
TENNESSEE
MISSISSIPPI
ALABAMA
GEORGIA
FLORIDA
S. CAROLINA
NORTH CAROLINA
VIRGINIA
W. VIRGINIA
MARYLAND
D.C.
DEL.
NEW JER.
PENNSYLVANIA
NEW YORK
VERMONT
NEW HAMP.
MASS.
CONN.
R.I.
MAINE
SCALE:
1 mm = 30 km = 19 mi
1 in = 480 mi = 772 km

Up and Down the Metric System

Use the figure to answer the conversion questions.

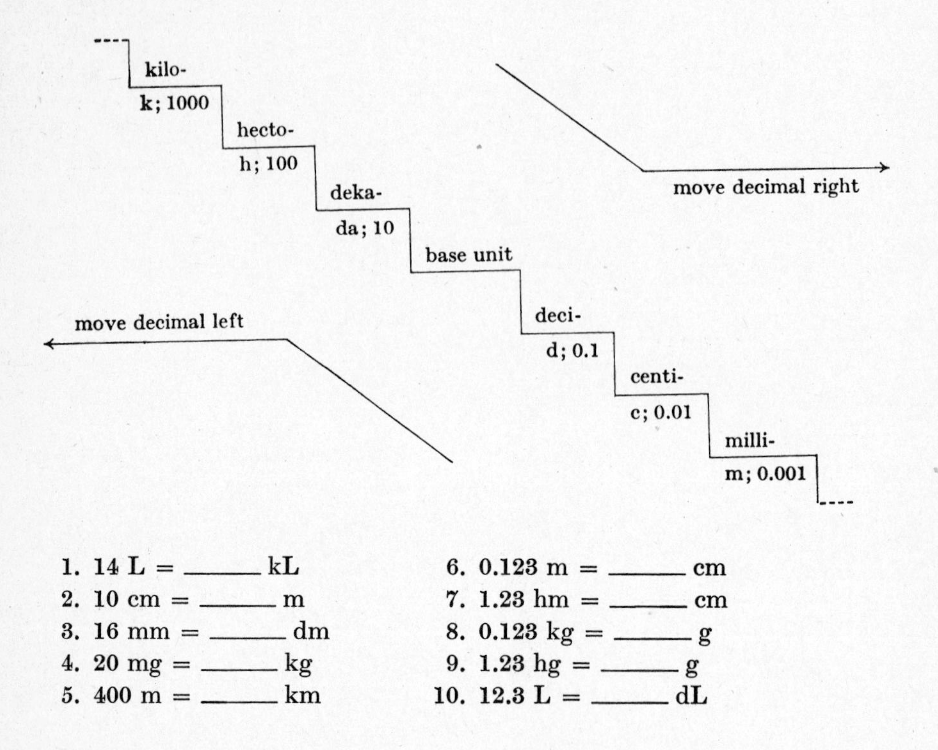

1. 14 L = ______ kL
2. 10 cm = ______ m
3. 16 mm = ______ dm
4. 20 mg = ______ kg
5. 400 m = ______ km
6. 0.123 m = ______ cm
7. 1.23 hm = ______ cm
8. 0.123 kg = ______ g
9. 1.23 hg = ______ g
10. 12.3 L = ______ dL

Color the Car

There are at least three variations to this activity.

1. Have children color squares at the bottom of the sheet each a different color. Cut out the squares. Match the squares to the corresponding squares in the picture. Color the congruent squares the same color.
2. Have children color squares at the bottom of the sheet each a different color. Have students measure a side of each square. Color the squares in the picture the same color as the corresponding squares at the bottom.
3. Have children color squares at the bottom of the sheet each a different color. Have students measure and find the area of each square. Color the squares in the picture the same color as the square at the bottom with the corresponding area.

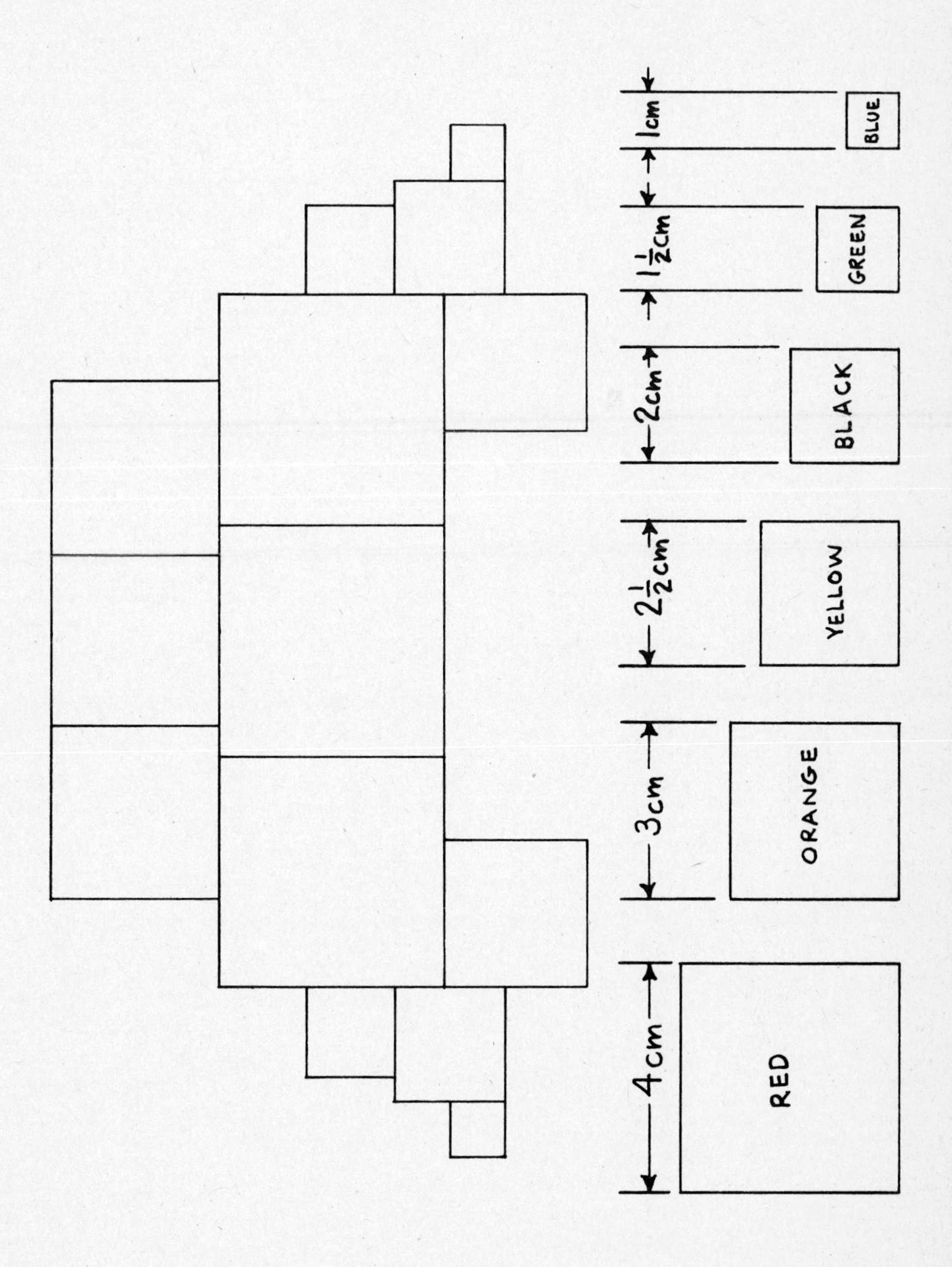
1cm
BLUE
$1\frac{1}{2}$cm
GREEN
2cm
BLACK
$2\frac{1}{2}$cm
YELLOW
3cm
ORANGE
4cm
RED

Metric Quiz

Consider the following statements and determine which are likely and which are unlikely. The number of answers you get correct will determine your metric rating as follows:

NUMBER CORRECT	METRIC RATING
0–3	A miserably messy metric measurer.
4–7	A moderately mediocre metric measurer.
8–10	A miraculously meticulous metric measurer.

1. The basketball player is three meters tall.
2. The bicycle was traveling 20 kilometers per hour.
3. He drank a liter of cola in one gulp.
4. The temperature dropped to 25° Celsius and it started to snow.
5. The football player weighed 120 kilograms.
6. The pencil weighed 100 grams.
7. His foot is five decimeters long.
8. The area of a postage stamp is 20 square centimeters.
9. He purchased 250 milliliters of cola for 15 cents.
10. The penny weighed about 3 grams.

METRIC QUIZ ANSWERS

1. Unlikely
2. Likely
3. Unlikely
4. Unlikely
5. Likely
6. Unlikely
7. Unlikely
8. Unlikely
9. Likely
10. Likely

APPENDIX A

Metric Resources

METRIC ARTICLES

"An American Dilemma: Measuring Up in the Future." *School Science and Mathematics*. May, 1971. Vol. 71, pp. 435–36.
"Adult Education and the Metric System." *Adult Education*. November, 1971. Vol. 20, p. 190.
"Changing to the Metric System: An Idea Whose Time Has Come." *NASSP Bulletin*. November, 1972. Vol. 56, pp. 47–59.
"The Coming of the Metric System: Bilingualism in Measurement." *Arithmetic Teacher*. May, 1973. pp. 397–99.
"Descriptive Analysis of the Teaching of the Metric System in the Secondary Schools." *Science Education*. February, 1969. pp. 89–94.
"Education in Decimal Currency and the Metric System." *School Science and Mathematics*. October, 1969. Vol. 69, No. 7, pp. 644–46.
"Experience, Key to Metric Unit Conversion." *Science Teacher*. November, 1970. Vol. 37, No. 8, pp. 69–70.
"Experiences for Metric Missionaries." *Arithmetic Teacher*. April, 1973. pp. 269–73.
"From a Panelist's Perspective." *Arithmetic Teacher*. April, 1973. pp. 245–46.
"Get Ready for the Metric System." *Instructor Magazine*. December 1971.
"Going Metric in Hawaii." *Arithmetic Teacher*. April, 1973. pp. 258–60.
"Ideas." *Arithmetic Teacher*. April, 1973. pp. 280–88.
"Inching Our Way Toward the Metric System." *Arithmetic Teacher*. April, 1973. pp. 275–79.
"The International System of Units (SI)." *Physics Teacher*. October, 1971. pp. 379–81.
"Is Metric a Measure of Pain?" *Industrial Arts and Vocational Education*. February, 1972. Vol. 61, pp. 22–24.
"Let's Start Measuring Up to the Metric Scale." *Nation's Schools*. November, 1971. Vol. 88, p. 26.
"Meaningful Metric." *School Science and Mathematics*. May 1964. Vol. 64, pp. 421–22.
"Measurement Standards, Physical Constants, and Science Teaching." *Science Teacher*. November, 1971. Vol. 38, pp. 63–71.
"Metric Is Here; So Let's Get on With It." *Arithmetic Teacher*. May, 1973. pp. 400–402.
"The Metric System." *Grade Teacher*. December, 1971.
"Metric System Ahead." *NEA Research Bulletin*. December, 1971. Vol. 49, pp. 109–12.

"The Metric System in the Elementary Grades." *Arithmetic Teacher*. May, 1967. pp. 349–53.

"The Metric System in Grade 6." *Arithmetic Teacher*, January, 1964. pp. 36–38.

"The Metric System: Its Relation to Mathematics and Industry." *Mathematics Teacher*. September, 1920. pp. 25–35.

"The Metric System in Junior High School." *Mathematics Teacher*. December, 1958. pp. 621–23.

"The Metric System—Let's Emphasize Its Use in Mathematics." *Arithmetic Teacher*. May, 1973. pp. 395–96.

"The Metric System: Past, Present-Future?" *Arithmetic Teacher*. April, 1973. pp. 247–55.

"The Metric System of Weights and Measures." *National Council of Teachers of Math*. 20th Year Book, 1948.

"Metrication in Britain." *Arithmetic Teacher*. April, 1973. pp. 261–64.

"Metrication in the School Curriculum." *Trends in Education*. April, 1972. Vol. 26, pp. 35–40.

"Metrication Urged by N.S.T.A. Committee." *Science Teacher*. January, 1971. Vol. 38, pp. 6–7.

"The Metrics Are Coming." *Grade Teacher*. February, 1971.

"New Dimensions for Practically Everything: Metrication." *American Education*. April, 1972. Vol. 8, pp. 10–14.

ORGANIZATIONS MARKETING METRIC MATERIALS FOR EDUCATORS

Activity Resources Co., Inc.
Box 4875
Hayward, CA 94545

Allyn & Bacon, Inc.
470 Atlantic Ave.
Boston, MA 02210

American Ass'n. of School Librarians
50 E. Huron St.
Chicago, IL 60611

American Nat'l. Standards Inst.
1430 Broadway
New York, NY 10018

American Nat'l. Standards Inst.
Central Instrument Co.
900 Riverside Dr.
New York, NY 10032

A. Balla & Co.
3494 N. Ocean Blvd.
Ft. Lauderdale, FL 33308

Behavioral Research Laboratory
Box 577
Palo Alto, CA 94302

Channing L. Bete Co., Inc.
45 Federal St.
Greenfield, MA 01301

Chas. E. Merrill Pub. Co.
1300 Alum Creek Drive
Columbus, OH 43216

Compu-Data Services, Inc.
16 Sherman St.
P.O. Box 471
Wayne, NJ 07470

Creative Publications
P.O. Box 10328
Palo Alto, CA 94303

C-Thru Ruler Co.
6 Britton Dr.
Bloomfield, CT 06002

Cuisenaire Co. of America
12 Church St.
New Rochelle, NY 10805

Dick Blick Co.
P.O. Box 1267
Galesburg, IL 61401

Dominie Pty. Ltd.
8 Cross St.
Brookvale, Australia 2100

Edmund Scientific Co.
380 Edscorp Bldg.
Barrington, NJ 08007

Educational Teaching Aids
159 W. Kinzie St.
Chicago, IL 60610

Enrich, Dept. M.
760 Kifer Rd.
Sunnyvale, CA 94086

Grolier Educational Corp.
845 Third Ave.
New York, NY 10022

Ideal School Supply Co.
11000 S. Lavergne Ave.
Oak Lawn, IL 60453

International Business Machines
Old Orchard Rd.
Armonk, NY 10504

International Tutors
Dept. A., 22303 Devonshire
Chatsworth, CA 91311

Jem Innovations
4568 E. 45th St.
Tulsa, OK 74135

John Colburn Assoc., Inc.
P.O. Box 187
Lake Bluff, IL 60044

John Wiley & Sons, Inc.
605 Third Ave.
New York, NY 10016

Laidlaw Brothers
Thatcher & Madison Ave.
River Forest, IL 60305

Larry Harkness Co.
115 N. Princeton Ave.
Villa Park, IL 60181

Leicestershire Learning Systems
Box 335
New Gloucester, ME 04260

MacLean-Hunter Learning Materials
481 University Ave.
Toronto 101, Ontario, Canada

Math Shop
5 Bridge St.
Watertown, MA 02172

Metric Ass'n., Inc.
Sugarloaf Star Route
Boulder, CO 80302

Metrix Corp.
P.O. Box 19101
Orlando, FL 32814

Midwest Pub. Co., Inc.
P.O. Box 129
Troy, MI 48084

Mind/Matter Corp.
P.O. Box 345
Danbury, CT 06810

Moyer Vico Ltd.
25 Milvan Dr.
Weston, Ontario, Canada

National Council of Teachers
of Mathematics
1906 Association Dr.
Reston, VA 22091

National Microfilm Ass'n.
Suite 1101, 8728 Colesville
Silver Spring, MD 20910

National Science Teachers Ass'n.
1201 16th St., N.W.
Washington, DC 20030

National Textbook Co.
8259 Niles Center Rd.
Stokie, IL 60076

National Tool, Die & Precis.
Machining Ass'n.
9300 Livingston Rd.
Washington, DC 20022

Pathescope Educational Films
71 Weyman Ave.
New Rochelle, NY 10802

Pickett Industries
P.O. Box 1515
Santa Barbara, CA 03102

Polymetric Services
4600 Brewster Dr.
Tarzana, CA 91356

Random House
The School Division
Westminster, MD 21157

Realty Facts
P.O. Drawer 449
Warwick, NY 10990

Robie Sales Co.
2755 Woodshire Dr.
Hollywood, CA 90068

Sargent-Welch Scientific
7300 N. Linder Ave.
Skokie, IL 60076

Selective Educational Equip.
3 Bridge St.
Newton, MA 02195

Sigma Scientific, Inc.
P.O. Box 1302
Gainesville, FL 32601

Spectrum Ed. Supp., Inc.
9 Dohme Ave.
Toronto, Ontario, Canada

Sterling Publishing Co., Inc.
419 Park Ave. South
New York, NY 10016

Swani Publishing Co.
Box 248
Roscoe, IL 61073

Teach'em, Inc.
625 N. Michigan Ave.
Chicago, IL 60611

Union Carbide Corp.
Educational Aids Dept.
P.O. Box 363–B
Tuxedo, NY 10987

FROM THE SUPERINTENDENT OF DOCUMENTS

Brief History and Use of English and Metric Systems of Measurement with a Chart of Modernized Metric System. (Rev. 1970.) (4) SD Catalog No. C13.10:304A. 25¢.

Education, An Interim Report of the U.S. Metric Study. National Bureau of Standards Special Publication. July, 1971. SD Catalog No. C13.10:345–6. $1.75.

The International Metric System of Weights and Measures. May 26, 1932. U.S. Department of Commerce.

A Metric America: A Decision Whose Time Has Come. National Bureau of Standards Special Publication 345. July, 1971. SD Catalog No. C13.10:345. $2.25.

Modern Weights and Measures Systems of Weights and Measures. U.S. Dept of Commerce.

Successful Experiences in Teaching Metric, ed. Jeffrey V. Odom. National Bureau of Standards Special Publication 441, January 1976. SD Catalog No. C13.10:441. $2.30.

Units of Weight and Measures: International Metric and U.S. Customary Definitions and Tables of Equivalents. National Bureau of Standards Miscellaneous Publication 286. SD Catalog No. 12.10:286. $2.25.

METRICATION AIDS

Classroom Metric Lines (Posters). Instructor Curriculum Materials, The Instructor Publications, Inc. Dansville, N.Y. 14437.

Introducing the Metric System. Programmed Learning Booklet. Coronet, 65 E. South Water St., Chicago, Ill. 60601. Grades 6–9.

Measurement Skills (kit). (English or Bilingual) (Spanish-English). Encyclopaedia Britannica, 425 N. Michigan Ave., Chicago, Ill. 60611.

Metric Conversion Card. National Bureau of Standards, Washington, D.C. 20234. Special Publication 365, issued July, 1972. SD Catalog No. C 13, 10:365. 10¢.

Metric Aids: A Catalog of Materials for Metric Teaching Aids. Metric Aids, Ltd., 75 Horner Ave., Toronto 530, Ontario, Canada.

"The Metric System." Study Prints. *Instructor Magazine*, 1971. Dansville, New York 14437.

Cuisenaire Rods. Useful for helping children in elementary grades visualize the logic and efficiency of the metric system. Cuisenaire Company of America, 12 Church St., New Rochelle, N.Y. 10805.

Metres, Litres & Grams. Schools Council. A pamphlet of value to curriculum workers responsible for developing metrication materials. Citation Press, 50 W. 44th St., New York, N.Y. 10036. $1.35.

Metric Training Aids (Price List). Metric Association, Inc. Sugarloaf Star Route, Boulder, Colo. 80302.

Metrication Posters. Polymetric Services, 4600 Brewster Dr., Tarzana, Calif. 91356.

Modernized Metric Systems Wall Chart. National Bureau of Standards, Washington, D.C. 20234. Special Publication 304 SD Catalog C 13, 10:304. 35¢.

Organizations Marketing Metric Materials for Educators. National Council of Teachers of Mathematics, 1906 Association Dr., Reston, Virginia 22091.

The Metric System: A Programmed Approach. Charles E. Merrill Pub. Company, 1300 Alum Creek Dr., Columbus, Ohio 43216. $2.95.

FILMS AND FILMSTRIPS

Area and Volume Measurement from Metric System. 16 mm. Sterling Pub. Co., Inc., 419 Park Ave. South, NY 10016. 1971. Grades 3–8.

Linear Measurement. 16 mm. Sterling Publ. Co., Inc., 419 Park Ave. South, NY 10016. 1971 Grades 3–6.

Man Is the Measure. Sound film-strip. Ford Motor Co., Detroit, MI. Grades 5–7.

Measurement Skills: Centimeters and Decimals. Filmstrip with cassette, workbook. Encyclopaedia Britannica, 425 N. Michigan Ave., Chicago, IL 60611.

Metric System. 13 min., black and white, 16 mm. McGraw-Hill, 330 W. 42nd St. NY 10036.

The Metric System of Measuring. Filmstrip. Encyclopaedia Britannica, 425 N. Michigan Ave., Chicago, IL 60611.

The Metric System: Universal Language of Measurement. 6 color filmstrips; 6 LP records. Pathoscope, 71 Weyman Ave., New Rochelle, NY 10802

Understanding and Using the Metric System. 80 slides Grades 5–7 Denoyer Geppert, 5235 Ravenswood Ave., Chicago, IL 60640.

Weighing and Measuring. 16 mm. Sterling Pub. Co., Inc., 419 Park Ave. South, New York 10016. 1971. Grades 3–8.

SOURCES FOR FILMS AND FILMSTRIPS

BFA Educational Media
2211 Michigan Ave.
Santa Monica, CA 90404.

Coronet Films
65 E. South Water St.
Chicago, IL 60601.

Eye Gate House, Inc.
146–01 Archer Ave.
Jamaica, NY 11435.

Ideal School Supply Co.
11000 S. Lavergne Ave.
Oaklawn, IL 60453.

Library Filmstrips Center
3033 Aloma
Wichita, KS 67211

NBC Educational Enterprises
30 Rockefeller Plaza
New York, NY 10020.

Society for Visual Education, Inc. 1345 Diversy Parkway
Chicago, IL 60614.

APPENDIX B

Metric Relations

PREFIX	SYMBOL	MULTIPLICATION FACTOR
tera	T	1 000 000 000 000 = 10^{12} (trillions)
giga	G	1 000 000 000 = 10^{9} (billions)
mega	M	1 000 000 = 10^{6} (millions)
kilo	k	1 000 = 10^{3} (thousands)
hecto	h	100 = 10^{2} (hundreds)
deka	da	10 = 10^{1} (tens)
	m, g, L; basic unit	1 = 10^{0} (ones)
deci	d	0.1 = 10^{-1} (tenths)
centi	c	0.01 = 10^{-2} (hundredths)
milli	m	0.001 = 10^{-3} (thousandths)
micro	μ	0.000 001 = 10^{-6} (millionths)
nano	n	0.000 000 001 = 10^{-9} (billionths)
pico	p	0.000 000 000 001 = 10^{-12} (trillionths)

APPENDIX C

Glossary of Symbols

A ampere. Base unit of electric current in both the SI and in the customary system.

°C Celsius. Alternate of the kelvin, the base unit of temperature in the SI. Degrees Celsius were formerly called *degrees centigrade.*

cd candela. Base unit of luminous intensity in both the SI and in the customary system.

cg centigram (10^{-2} gram.) One-hundredth of a gram or 10 milligrams.

cL centiliter (10^{-2} liter). One-hundredth of a liter or 10 milliliters.

cm centimeter (10^{-2} meter). One-hundredth of a meter or 10 millimeters.

cm^2 square centimeter. The area of a square whose sides are 1 cm in length.

cm^3 cubic centimeter. The volume of a cube whose sides are 1 cm in length.

dag dekagram (10^1 grams). Ten grams; approximate weight of two nickels.

daL dekaliter (10^1 liters). Ten liters; about 2.6 gallons.

dam dekameter (10^1 meters). Ten meters; about 11 yards.

dg decigram (10^{-1} gram). One-tenth of a gram or 10 centigrams.

dL deciliter (10^{-1} liter). One-tenth of a liter or 10 centiliters.

dm decimeter (10^{-1} meter). One-tenth of a meter or 10 centimeters.

g gram. A metric unit of mass or weight equal to 1/1000 kilogram; nearly equal to the weight of 1 cubic centimeter of water at its maximum density.

ha hectare (10^4 square meters). The area of a square whose sides are 100 m in length; will replace the acre in area usage.

hg hectogram (10^2 grams). One hundred grams.

hL hectoliter (10^2 liters). One hundred liters.

hm hectometer (10^2 meters). One hundred meters.

K kelvin. The unit of temperature measurement of the SI; for normal use expressed in *degrees Celsius.*

kg kilogram (10^3 grams). One thousand grams; commonly called a *kilo.*

kL kiloliter (10^3 liters). One thousand liters.

km kilometer (10^3 meters). One thousand meters.

L liter. A metric unit of liquid capacity; equal to 1 cubic decimeter (dm^3).

m meter. Base unit of length in the SI; equal to approximately 1.1 yards.

m^2 square meter. The area of a square whose sides are 1 m in length.

m^3 cubic meter. The volume of a cube whose sides are 1 m in length.

mg milligram (10^{-3} gram). A small unit of weight equal to 1/1000 of a gram.

mL milliliter (10^{-3} liter). A small unit of volume equal to 1/1000 of a liter.

mm millimeter (10^{-3} meter). A small unit of length equal to 1/1000 of a meter.

mol mole. The base unit of amount of substance in the SI.

NBS	National Bureau of Standards, Washington, D.C.
s	second. The base unit of time in the SI and the customary system.
SI	*Le Système International d'Unités.* The International System of Units, which is used worldwide; commonly referred to as the *metric system.*
t	metric ton. Measure of weight equal to 1000 kilograms or about 2200 pounds; may be spelled *tonne.*

APPENDIX D

Additional Metric Definitions

Time: second (*s*). The second is defined as the duration of 9,192,631,770 cycles of the radiation associated with a specified transition of the cessium-133 atom. It is realized by tuning an oscillator to the resonance frequency of cesium-133 atoms as they pass through a system of magnets into a detector.

Electric Current: ampere (*A*). The ampere is defined as that current which, if maintained in each of two long parallel wires separated by one meter in free space, would produce a force between the two wires (due to their magnetic fields) of 2×10^{-7} newton for each meter of length.

Temperature: kelvin (*K*). The kelvin is defined as the fraction 1/273.16 of the thermodynamic temperature of the triple point of water. The temperature 0 K is called "absolute zero." On the commonly used Celsius temperature scale, water freezes at about 0 °C and boils at about 100 °C. The °C is defined as an interval of 1 K, and the Celsius temperature 0 °C is defined as 273.15 K. 1.8 Fahrenheit degrees are equal to 1.0 °C or 1.0 K; the Fahrenheit scale uses 32 °F as a temperature corresponding to 0 °C.

Amount of substance: mole (*mol*). The mole is the amount of substance of a system that contains as many elementary entities as there are atoms in 0.012 kilogram of carbon-12. When the mole is used, the elementary entities must be specific and may be atoms, molecules, ions, electrons, other particles, or specified groups of such particles.

Luminous intensity: candela (*cd*). The candela is defined as the luminous intensity of 1/600 000 of a square meter of a blackbody at the temperature of freezing platinum (2045 K).

APPENDIX E

Common Conversions

Length (Approximates)

1 inch = 2.54 centimeters	1 centimeter = 0.39 inch
1 foot = 0.30 meter	1 meter = 39.37 inches
1 yard = 0.91 meter	1 meter = 1.09 yards
1 mile = 1.61 kilometers	1 kilometer = 0.62 mile

Weight (Approximates)

1 ounce = 28.35 grams	1 gram = 0.035 ounce
1 pound = 0.45 kilogram	1 kilogram = 2.21 pounds

Volume (Approximates)

1 liquid quart = 0.95 liter	1 liter = 1.06 liquid quarts
1 liquid gallon = 3.79 liters	1 liter = 0.26 liquid gallon

Area (Approximates)

1 hectare = 1 square hectometer = 10 000 square meters = 2.5 acres

Temperature

Celsius Degrees = 5/9 (Fahrenheit − 32°)

Fahrenheit Degrees = (9/5 Celsius) + 32°

Celsius	Fahrenheit	
−40°	−40°	
−20°	−4°	
−10°	14°	
0°	32°	water freezes
10°	50°	
20°	68°	recommended room temperature
30°	86°	
37°	98.6°	body temperature
40°	104°	very hot day
50°	122°	
100°	212°	water boils

APPENDIX F

U.S. DEPARTMENT OF COMMERCE
National Bureau of Standards
Washington, D.C. 20234

Letter Circular 1052
February 1974

All You Will Need to Know About Metric
(For Your Everyday Life)

Metric is based on Decimal system

The metric system is simple to learn. For use in your everyday life you will need to know only ten units. You will also need to get used to a few new temperatures. Of course, there are other units which most persons will not need to learn. There are even some metric units with which you are already familiar: those for time and electricity are the same as you use now.

BASIC UNITS

METER: a little longer than a yard (about 1.1 yards)
LITER: a little larger than a quart (about 1.06 quarts)
GRAM: a little more than the weight of a paper clip

(comparative sizes are shown)

25 DEGREES FAHRENHEIT

25 DEGREES CELSIUS

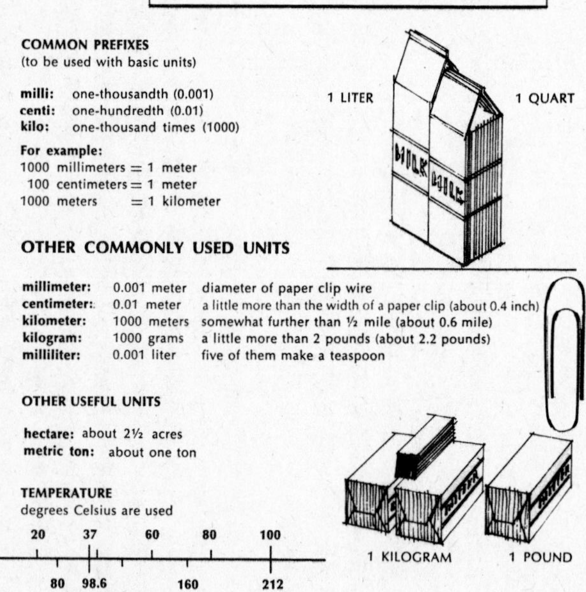

COMMON PREFIXES
(to be used with basic units)

milli: one-thousandth (0.001)
centi: one-hundredth (0.01)
kilo: one-thousand times (1000)

For example:
1000 millimeters = 1 meter
100 centimeters = 1 meter
1000 meters = 1 kilometer

OTHER COMMONLY USED UNITS

millimeter:	0.001 meter	diameter of paper clip wire
centimeter:	0.01 meter	a little more than the width of a paper clip (about 0.4 inch)
kilometer:	1000 meters	somewhat further than ½ mile (about 0.6 mile)
kilogram:	1000 grams	a little more than 2 pounds (about 2.2 pounds)
milliliter:	0.001 liter	five of them make a teaspoon

OTHER USEFUL UNITS

hectare: about 2½ acres
metric ton: about one ton

TEMPERATURE
degrees Celsius are used

°C	−40	−20	0	20	37	60	80	100
°F	−40	0	32	80	98.6		160	212
			water freezes		body temperature			water boils